FARM TRACTORS

THE HISTORY OF THE TRACTOR

ROBERT N. PRIPPS

PHOTOGRAPHY BY ANDREW MORLAND

Voyageur Press

Published in 2011 by Voyageur Press, an imprint of MBI Publishing Company, 400 First Avenue North, Suite 300, Minneapolis, MN 55401 USA

Voyageur Press titles are also available at discounts in bulk quantity for industrial or sales-promotional use. For details write to Special Sales Manager at MBI Publishing Company, 400 First Avenue North, Suite 300, Minneapolis, MN 55401 USA.

To find out more about our books, visit us online at www.voyageurpress.com.

ISBN-13: 978-0-7603-4051-6

Editor: Michael Dregni
Designed by: Andrea Rud
Cover designed by: Simon Larkin

Printed in China

On the front cover:
Main image: A 1948 Farmall HV, owned by Larry Maasdam.
Insets, left to right: A 1915 Hart-Parr 50/60 "Old Reliable," owned by Gary Spitznogle; a 1952 John Deere Model AWH, owned by Larry Massdam; and a 1979 Ford FW-50, owned by Dale Bissen.

On the endsheets: A threshing crew poses proudly with its Rumely OilPull kerosene tractor and separator. *Glenbow Archives*

On the title pages: The VAO orchard version of the popular Case VA Series was built from 1942 to 1955; this is a 1947 model.

On the back cover:
Top: John Deere advertising artwork.
Center: A 1934 Caterpillar Diesel Forty, owned by Alan Smith.
Bottom: A 1955 John Deere Model B-Garden tractor, owned by Walter Keller.
Right: A 1950s Massey-Harris brochure.

Acknowledgements

As might be expected, I received much help in writing this book. First, I must credit Michael Dregni, editorial director for Voyageur Press, for it was his idea to put this book together. He and the staff at Voyageur deserve credit for making the book look and read as well as it does. My thanks to them.

This is primarily a picture book. My thanks to Andrew Morland for these excellent photos.

There were many who pushed, washed, started, moved, and struggled with their antiques so that Andrew Morland could get the photos that grace the following pages. They also provided reams of data and historical details. I won't list them here for fear of missing and offending someone.

I will list several, however, who went way out of their way to help us, often giving us a meal or two, as well as days out of their otherwise busy schedules: Les Abraham, Heritage Hall Museum, Owatonna, Minnesota; Tom Armstrong, N-Complete Remanufactured Ford Tractors; Glen Braun; Keith Bruder; Alan Buckert; Gary Burkey; Karen Chabal; Ed Claessen; Milan Duchaj; Great Dorset

Steam Fair, Dorset, Great Britain; Carl Halverson; the Rich Holicky family; Bill Karl; Ken Kass; the Walter Kellor family; Larry and Ryan Maasdam; Midwest Old Threshers; Lennis Moore; the Eldon Oleson family; Ken and Dan Peterman; David Pruehs; Gary and Marlin Spitznogel; Frank Sticha; Mike Thorne; Wayne Timm; Steve Tolander; Billie Turley; and the Vouk family. My thanks to all.

Robert N. Pripps
Springstead, Wisconsin

1927 Huber 25/50
The 25/50 was a big tractor in any language, weighing 5 tons (4,500 kg). It was produced by the Huber Manufacturing Company of Marion, Ohio, from 1927 to 1941. After it was tested in 1927 by the University of Nebraska, subsequent production versions were labeled 40/62. The tractor featured a Sterns four-cylinder engine and two-speed transmission. Owner: the Timm family.

Contents

1907 Minneapolis 45

1919 Holt 10-Ton

1919 Wallis Model K

1929 Minneapolis 27/42

1938 Minneapolis-Moline UDLX

1999 Case-IH Steiger 9330

Introduction

"Many shall run to and fro, and knowledge shall be increased."
—Daniel 12:4

For three weeks in June 1999, we toured the American Midwest, getting words and pictures for *The Big Book of Farm Tractors.* We avoided freeways and chose the country roads—the blue highways on our map. Likewise, this book is not the usual straight history with chapters about each brand, but is instead a meander through the backroads of power farming from the early days to the modern era.

This journey begins with the great steam engines before the American Civil War. It then travels through the tentative beginnings of the internal-combustion era, through two world wars and the Great Depression, and on into the era of the diesel engine. There are peaks and valleys, ups and downs in the industry. We will see most of the big names disappear along the way. We will also see perseverance and ingenuity pay off in a big way for some.

This admittedly is history painted with a broad brush. It was our goal, however, to get into the personal side of the history of tractors and how the tractor fit into and influenced the life and times of the farm during the past 150 years. This is the story of people, both on the farms and in the great industries. Vignettes, quotes, timelines, and sidebars will help to fill in the gaps and indicate to the reader what else was going on in the country at the same time.

The practice of agriculture changed little from the beginning of time to the early nineteenth century. Then, changes came rapidly. Those of you who are, or were, farmers will probably see yourselves in the following pages. Hopefully, those readers with no connection to actual farming will gain a new perspective of the wrenching changes that have taken place on the farm over the years. In either case, we hope you'll enjoy this trip through power-farming history.

Robert N. Pripps
Andrew Morland

1934 Caterpillar Diesel Forty
The Diesel Forty was an upgraded version of the Diesel Thirty-Five, which was sold in 1933 and 1934. Tested by the University of Nebraska in 1935, the Diesel Forty tipped the scales at 15,642 pounds (7,039 kg). Owner: Alan Smith.

1938 Minneapolis-Moline UDLX
Only about 150 of the stylish, but not truly practical, Model U-Deluxe tractors were made by the Minneapolis-Moline Company of Minneapolis, in the years 1938 through 1941. Today they are among the most desirable and sought after of all vintage tractors.

Prologue

"It all begun on a midsummer day in 1831 near Steel's Tavern, Virginia. The stillness of the countryside was about to be broken by a public demonstration that would mark the beginning of a new epoch in agricultural invention. The small crowd of bystanders was curious and skeptical, but as Cyrus H. McCormick's new creation moved down the field the wheat fell in a steady stream upon its platform. The whirling gears of the mechanical reaper soon would be a familiar sound in the American harvest field."
—Marvin McKinley,
Wheels of Farm Progress, 1980

Cyrus Hall McCormick had no idea of the changes he was unleashing on that midsummer day in 1831, either for himself or for agriculture. In fact, he thought of his "reaper" as a possible labor-saving device for his own farm only.

Prior to McCormick's reaper, the handheld cradle scythe was the tool for grain cutting. Scything was extremely difficult and skilled work, and five acres (2 hectares) of cut grain per person per day was the accepted rate. A cradler was a specialist, commanding as much as three times the daily pay rate of other farmhands. With McCormick's reaper, a farmer and a horse could harvest ten acres (4 hectares) per day, although separate farmhands to serve as rakers, bundlers, and shockers were still needed.

It wasn't until Obed Hussey of Baltimore, Maryland, applied for a patent on his own reaper invention in 1833 that McCormick recognized the implications and applied for his own patent. Citing prior art, McCormick challenged Hussey in the courts and in the fields of North America. McCormick's reaper was the first of a series of mechanical revolutions in the farm fields of the 1800s that led the way for the farm tractor, inventions that had a profound impact on not only agriculture but on our culture as a whole.

John F. Appleby's amazing Appleby Automatic Knotter of 1875 revolutionized the grain reaper, turning it into the labor-efficient binder and suddenly saving thousands of man-hours per harvest. The binder, cut the grain, sheaved it, and tied it with a cord still known today as "binder twine." The tied sheaves were deposited in the field, ready for shocking. The binder was the first mechanically complicated machine that confronted the farmer, who had to learn to use it, or else!

The rapid increase in acreage harvested per individual worker brought on by McCormick and Appleby's inventions mothered the next big step—the threshing machine, or separator. Previously, threshing had been done with a flail, which was also skilled work similar to scything grain. A good flail thresher could only process about seven bushels of wheat in a long day. Now, the threshing process had to be mechanized to keep up with the harvesting work done by the binder.

The Jacob Pope's Groundhog thresher was the first on the North American scene, in 1802. Since it was invented before the reaper, its purpose was not necessarily to speed up the process, but to improve efficiency over the flail. Nevertheless, the hand-cranked Groundhog bore the elements of subsequent high-volume machines.

Pope's Groundhog only dehulled the grain, however. In 1837, Hiram A. Pitts created his Buffalo Pitts thresher and cleaner that included a fanning mill

1860s Fearless thresher and power advertisement

Small firms throughout North America built a variety of separators, including the Fearless thresher made by the Empire Agricultural Works of Cobleskill, New York. Power, in the days before the widespread use of the steam engine, came from true horse power. The Fearless power unit was a tread machine for a two-horse team that fed power through a governor to a belt drive.

McCormick harvester at work, 1831

Hat's were raised to the successful first demonstration of Cyrus Hall McCormick's harvester on a midsummer day in 1831. The McCormick harvester became the foundation of the massive and supremely influential International Harvester Company of Chicago.

and walker to blow out chaff and carry out straw. The Buffalo Pitts machine required more power to operate, however. So, a horse tread was added to give one, two, or three "horsepower." When still more power was needed, horse sweeps came into being with as many as eighteen horses. With such a rig, a threshing rate of fifty to sixty bushels an hour was possible.

By the time of the American Civil War, the threshing machine had developed into mature technology. It was then made of wood, which required fine craftsmanship in its construction. It still lacked the feeders and straw stackers that would be added in the last half of the 1800s. What was needed now was steady, untiring power; the more power, the wider the threshing mechanism that could be accommodated.

And that's where the steam engine enters the North American farm scene.

1880s McCormick binder advertisement

The addition of the twine-binding mechanism as created by John F. Appleby suddenly revolutionized harvesting and reduced the amount of workers needed to cut wheat and other crops.

CHAPTER 1

The Steam Era, 1855–1920

"I foresee the successful application of steam power to farm work."
—Abraham Lincoln,
speech at the Wisconsin State Fair, 1859

1900s Nichols & Shepard 20/70
Main photo: *Running on full throttle, steam fills the air and the belt purrs on this Nichols & Shepard 20/70. Known as the famous Red River line, Nichols & Shepard's steam engines and threshers were made in Battle Creek, Michigan.*

Keck-Gonnerman trademark
Above: *The trademark on a Keck-Gonnerman steam engine shows the home of a prosperous thresherman. Keck-Gonnerman of Mt. Vernon, Indiana, began business in 1873, making steamers, threshers, and sawmills. The firm went into gas tractors in 1917 and went out of business after World War II.*

The Evolution of Steam Power

STILLWATER ENGINE.

1880s Stillwater engine

In the early days of steam, engines were typically known as "thresher engines" as they were designed simply to power threshing machines. The first steamers were basically stationary engines on wheels and were pulled into place by a horse team. This later Stillwater steamer was self propelled, with the front wheels steered by chain. It was built by the venerable Northwest Thresher Company of Stillwater, Minnesota.

Timeline

1850: California becomes a state

1851: The sewing machine is invented simultaneously by Singer, Howe, and Hunt

1852: The first successful airship is flown by Henri Giffard of France

1861: Abraham Lincoln becomes U.S. President

1861–1865: American Civil War

1862: The Homestead Act gives 160 acres to American frontier settlers

1863: Lincoln delivers the Gettysburg Address

1876: Alexander Graham Bell invents the telephone

1877: Thomas Alva Edison invents the phonograph

1879: Edison invents the incandescent light

1903: Ford Motor Company established

1914: World War I begins

1917: Communists take over Russia

1920: U.S. women win the right to vote

Confining the vapors of boiling water causes a dramatic buildup of pressure. Expanding these vapors through pistons and cylinders, turbines, or even jets results in the potential for useful work to be done.

Hero of Alexandria, Egypt, born about 130 B.C., was the first to recognize the ability of heated air or water to perform tasks. His "air engine" used air pressure generated by heating air in a closed container to force water from one jar into another. As the second jar filled and became heavier, it pulled open temple doors through a series of ropes and pulleys. When the fire went out, the water siphoned back, allowing a counterweight to close the doors. Naturally, the fire that activated the temple doors was the fire on the altar.

Hero's "Aeolipile" steam engine was also the first known reaction, or jet, engine. This steam engine did no practical work but was made to demonstrate a principle. It consisted of a sphere into which steam was fed through hollow support tubes. The steam escaped through two bent tubes on opposite sides of the sphere. Reaction to the escaping steam's force caused the sphere to rotate.

In the subsequent years, many attempts were made to harness the force of steam by means of a piston in a cylinder. Low-pressure steam was used because the materials available would not tolerate higher pressures. A pressure of only 1 pound per square inch (psi) above atmospheric was used by Englishman Thomas Newcomen in a steam-powered mine pump in 1712. The engine consisted of a vertical brass cylinder and piston, with the top of the cylinder open to the air. Steam from a boiler was fed into the cylinder under the piston and was then cooled and condensed back to liquid by injecting cold water into the cylinder. As the steam condensed, it occupied less space and a vacuum was created beneath the piston. Atmospheric pressure then acting on the piston forced the piston down on its power stroke. The piston was connected to a rocking beam; the beam's other end operated a mine-water lift pump. The pump mechanism's weight pulled the piston back to the top of the cylinder for another cycle. The engine, which had automatic control valves, could make eight or ten strokes per minute.

Scottish inventor James Watt developed the basic Newcomen engine into a more efficient concept. Patented in 1769, his engine employed a separate condensing chamber into which the steam was discharged and converted back to water. Rather than injecting cold water into the cylinder, the cooling was done in the remote condenser, saving three-fourths of fuel costs.

Watt later devised the first crankshaft and flywheel. This allowed the steam piston, with its linear motion, to produce continuous rotary motion. Watt's contributions also included valve gearing, double-acting pistons, throttle controls, and the governor. Watt is frequently referred to as the inventor of the steam engine because these elements made it a practical device.

While Watt's engines employed steam or flywheel power to return the piston for the next power stroke, the power was still applied by atmospheric pressure pushing the piston into a vacuum created by condensing steam. Engines with up to 76-inch-diameter (190-cm) pistons were common. Later, Watt used steam pressure, just above atmospheric pressure, for the power stroke; metallurgy of the mid-eighteenth century was not sufficiently advanced to handle much higher pressures with safety. Also, the building of accurate cylinders was a problem. In about 1762, pioneer iron foundryman John Wilkinson invented a boring machine capable of making the cylinders that Watt needed: Watt once commented that a Wilkinson cylinder was out of round by only 0.375 inches (9.375 mm).

Watt's associate, William Murdock, built a crude rail locomotive on the principles of Watt's patents in 1784, but the first practical rail locomotive was built in 1804 by Englishman Richard Trevithick. Trevithick was the first to employ high-pressure steam, using about twice atmospheric pressure at first. American inventor Oliver Evans collaborated with Trevithick in his experiments, eventually operating an engine at 200 psi. The Trevithick engine had four, smooth driving wheels running on smooth rails, proving that sufficient traction could thus be obtained. An important concept in the Trevithick engine was that the exhausted steam was not condensed, but was released into the smokestack, providing forced firebox draft. This design has been used in virtually all subsequent steam engines and provides the characteristic *chuff-chuff* or *choo-choo* sound.

Railroads were a boom industry in the mid-nineteenth century. The newly created Janney safety car coupling and Westinghouse air brake linked trains together and allowed them to brake. By 1848, there were 6,000 miles (9,600 km) of tracks in North America with 2,000 miles (3,200 km) added annually.

The farm steam engine was a result of progress made in railroad engines. With the development of the thresher, the horse became insufficient as a power source. Hence, it was mostly the thresher manufacturers that began building portable steam engines for the farm: A. M. Archambault & Company of Philadelphia in 1849; Hoard & Bradford of Watertown, New York, in 1850; Gaar, Scott & Company of Richmond, Indiana, in 1852; M. Rumely Company of La Porte, Indiana, in 1863; and J. I. Case Threshing Machine Company of Racine, Wisconsin, in 1869. Most of these companies either built threshers based on patent rights purchased from Pitts, or of their own design.

1912 Reeves Canadian Special 32/120
The Reeves Company was founded in Columbus, Indiana, in 1874. It was taken over by the Emerson-Brantingham Implement Company in 1912 and moved to Rockford, Illinois, shortly thereafter. The old Reeves plant in Columbus later became the home of diesel engine builder Cummins.

Russell steamer, 1890s
A custom threshing crew stands proudly before its Russell steam engine and separator on their way to thresh grain in Modoc County, California. The Russell was built by Russell & Company of Massillon, Ohio.

1880s Pitts steamer advertisement

The famed Pitts Agricultural Works of Buffalo, New York, offered a variety of steam engines that could be fueled by coal, wood, or straw.

In California, the Daniel Best Agricultural Works of San Leandro, California, built its traveling combined harvester powered by as many as forty horses, with ground wheels driving the threshing and separating mechanisms. Almost anything could spook such a menagerie into an uncontrollable runaway. The ground-wheel power would then drive the mechanisms to destruction. Best sold his first steam tractor in 1889. It was followed—literally—by a steam-powered combine that same year. Steam from the engine was piped back through a high-pressure hose to power the combine mechanisms.

With the advent of steam power on the farm, horse-powered treadmills and rotary sweeps began to disappear from the scene. Steam, the first non-animal farm power source, was taking over—although there was much resistance to the change at first on the part of the farmer.

Farms in the late 1800s were much smaller on average than what is now considered a normal size; some families subsisted on as few as ten acres (4 ha), cleared. In those days, farming was mostly a self-sufficient occupation. Besides staples like wheat, corn, potatoes, and hay, sheep were raised for meat and wool, cows for milk products and meat, chickens for eggs and meat. Most farms featured fruit trees and a large vegetable garden. In the north, maple trees were tapped for syrup and sugar. Hunting, fishing, and trapping provided food and skins. The women canned, spun, sewed, baked, churned, and made candles and soap. Excess crops were sold to get the few necessities of life not produced on the farm. Naturally, a farmer would not want to risk thousands of dollars on a machine that could only be used at threshing time. And if it failed, where would the horsepower come from?

1907 Minneapolis 45
Left, both photos: *The Minneapolis Threshing Machine Company of Minneapolis, made steam engines from 1890 through 1924 that were single-cylinder, single-tandem-compound, and double-tandem-compound. This 45-hp, single-tandem-compound engine was only built in 1907, and this is one of just seven made. Steam exhausted from the 10.25-inch (25.625-cm) high-pressure cylinder to be utilized again in the 15-inch (37.50-cm) low-pressure cylinder. The stroke was 12 inches (30 cm). Owner: Norman Pross.*

Case

J. I. Case Threshing Machine Company
of Racine, Wisconsin

Jerome Increase Case's company built many of the most significant farm engines of the steam era. Case began experimenting with steam in the early 1860s after its Sweepstakes threshing machine was introduced. The Sweepstakes utilized horse-powers driven by as many as eighteen horses. About the same time, Case began using steam factory power.

The first Case-produced steam engine, called "Old No. 1," made its debut in 1869. It was not self-propelled, but pulled to the job by a team of horses. Once belted up, it produced about 8 hp.

Between 1869 and 1880, various inventions were devised for steering and propulsion of steamers. C. and G. Cooper of Mt. Vernon, Ohio, invented a bevel-gear set that could be added to most any portable engine for propulsion. Steering was still done by the team of horses, however.

Case added propulsion to its engines in 1878, but retained horse steering. Case's drive incorporated a ratchet-type differential. A 10-hp engine was added to the line that same year, with the same type of propulsion.

In 1880, Jessee Walrath was made manufacturing superintendent of Case. In those days in many factories, the person with that title had the responsibility for new designs and improvements. Walrath contributed mightily to the development of Case engines over the next sixteen years, his inventions including the straw-burner chute, the best steering mechanism in the industry, a full spur-gear differential, and a friction clutch. Models were offered over the years with side and center cranks, direct and reverse flues, chain and gear drives.

Case steamer and plow, 1900s
A plowing crew rests while breaking the prairie sod with their Case steamer.
(Glenbow Archives)

17

In 1887, Walrath devised a totally enclosed double cylinder featuring a rocker valve. This type of cylinder became known simply as the "Walrath." Use of the Walrath reduced the weight of a 12-hp engine by almost 3,000 pounds (1,350 kg).

George Morris became superintendent in 1897 when Walrath resigned. Morris also made continuous improvements in the engines over the years. He is most famous for putting springs between the boiler and wheels, allowing lighter engines, as the springs isolated the boiler from stresses and shocks when traveling on the roads. The unique and patented feature allowed the gears to stay in proper mesh while the drive axle moved up and down on the springs. By 1900, Case engines were outselling all competition.

As experience was gained and engine design matured, Case began experimenting with larger engines. In 1904, Case offered a 150-hp behemoth weighing 40,000 pounds (18,000 kg) that was obviously too big for field use. A few were built in the following years for road use before the design was abandoned.

Of greater importance was the famous Case 110, possibly the best and most popular steamer of the period. The Case 110 took more gold medals at the famed trials in Winnipeg, Manitoba, Canada, than any other engine.

In 1912, Case's steam engine production peaked at 2,250. After that, production diminished more rapidly than it had risen. By 1915, production dropped to 950 engines, and by 1920, only 346 were built. Production ceased in 1926 after fifty-seven years. Interestingly, the last engine built was a portable, like Case's Old No. 1.

1890s Case advertisement
"There are no others!" shouted this advertisement for Case's new spring-mounted steam engine.

The New Side Crank
SPRING MOUNTED ENGINE

Here It Is!   There are No Others!

The Agitator Separator
NOW, AS ALWAYS, SPEAKS FOR ITSELF.

Case Weighers, $50.   Case Loaders, $35.

J. I. CASE THRESHING MACHINE CO., Racine, Wis.

1890s Case steamer advertisement

Right: *A colorful advertisement for Case's steam engines "for farmers, contractors and municipalities."*

1913 Case 110

Below: *A 110-hp Case steamer blows out a plume of smoke as it pulls a twelve-bottom gang plow at a modern-day tractor show. The 110-hp boasted a 12x12-inch (300x300-mm), "simple-cylinder" engine. It weighed an amazing 18 tons (16,200 kg), complete with the fancy locomotive-style cab. Owners: Jim Briden and Norman Pross.*

Case steamer, 1941

Above, top: *Many Case steamers were still going strong decades after the company halted production. This steam engine was still at work in Montgomery, Minnesota, in 1941.*

1913 Case 80

Above, bottom: *This 80-hp Case utilized an 11x11-inch (275x275-mm), "simple-cylinder" engine and up to 150 psi of steam pressure.*

1917 Case 80

Above: *The Vouk family of St. Stephen, Minnesota, has owned this 1917 Case steamer since new. It was bought new by grandfather Frank Vouk when he was nineteen years old. He was a horse trader, sawmill operator, and custom thresherman, and put the Case to good use over the decades.*

1913 Case 110 logo

Right: *Case used this trademark transfer on its steam engines in 1913. It showed the Case factory on the Racine River in Racine, Wisconsin. At each end of the transfer are "Old Abe" eagles perched on globes, signifying the worldwide scope of Case by that time.*

1890s Daniel Best steamer advertisement

"A Revolution in Plowing," promised this ad. Labeling its machines "The Monarch of the Field," Daniel Best stated that his steamers "will do the work of 100 horses."

Holt steamer, 1900s

A Holt steamer hauls a train of three wagons laden with massive loads of cut boards from a sawmill in the Pacific Northwest. Ben Holt and his Holt Manufacturing Company of Stockton, California, followed the lead of arch-rival Daniel Best in building a steam traction engine. Both firms initially built their steamers to power their own separate advanced combined harvesters. (Eastman Collection, University of California–Davis)

Best
Daniel Best Agricultural Works of San Leandro, California

Daniel Best was born in Ohio in 1838. His family moved to Keokuk, Iowa, from where Daniel, then twenty-one, left for Oregon. After several years in the sawmill business in Oregon and Washington, he moved to California where he had three brothers farming large wheat ranches. While assisting his brothers harvest wheat, Daniel came up with the idea of a portable grain cleaner. Previously, the grain was hauled to a cleaning establishment, which was not only time consuming but also expensive. The following winter, he made three of his cleaning machines, which could be moved from farm to farm. The next fall, he and his brothers ran these machines for other farmers.

This was the start of a firm called Best Manufacturing Company. Cleaners were manufactured at first, but Daniel Best soon had developed a header-thresher-cleaner machine called a combined harvester. In fall 1886, he sold six of these to neighboring farmers. To power the combined harvester, or "combine" as they came to be called, a large team of draft animals was required.

An acquaintance of Best's from his Oregon days named Marquis DeLafayette Remington had patented a steam traction engine in 1888. The engine was unique in that it used a vertical boiler set far to the rear between the large drive wheels. A single front wheel was steered by steam power steering. The boiler arrangement reduced stresses on the pressurized components when traversing rough ground and made the machine lighter than others of its time. Remington's engine was designed primarily for traction work, with belt work a secondary consideration. Several of his engines were employed in the big western logging operations, where the Remington steamer was probably the first traction engine to be used to skid the giant logs.

Unfortunately, Remington's factory was nearly destroyed by a disastrous fire. To get cash to rebuild, he drove his remaining engine to Best's operation. After a brief demonstration, Best bought the patent rights to the engine for all but Oregon.

Best then set to work improving and enlarging the engine and integrating it with his combine. A steam hose from the engine provided power to operate the cutter, thresher, and cleaning mechanisms. It also powered a conveyor belt that brought straw forward to be used for fuel. Thus, the rig provided its own fuel as it went along.

The Best steamer was a great success with the large-acreage farmers of California, but the Best steamer still found most of its work in logging. By 1900, Best was turning out about twenty engines per year, some of which were being shipped to the far reaches of the globe.

Daniel Best steamer, 1894

Lumberjacks pause while hauling a gigantic load of monstrous logs behind their Daniel Best steamer. Best made fine use of his own name in advertising his machines as "The Best." Best's steamers burst onto the logging scene in the Pacific Northwest in the 1890s, revolutionizing the industry. In creating his steamer, Best bought the rights to the steam engine of DeLafayette Remington of Woodburn, Oregon, improved on the design, and went into production in 1889. The Remington was designed for use in the woods; the Best engines were suitable for logging as well as for farm use. (Eastman Collection, University of California–Davis)

The Romance of Steam

1900s Lion steamer
Above: *Threshing machines and steam engines of the Canadian Waterloo Manufacturing Company were sold as the Lion brand from 1850 to 1925. This trademark transfer graced a 16-hp engine. Owner: Ontario Agricultural Museum.*

To many a young farm boy or girl, the traveling steam engineer was the icon of achievement. The grease-covered, tobacco-chewing, well-traveled boss of the crew held the highest fascination for the rural lad or lass, who had probably not been more than twenty miles from his or her birthplace, nor handled more power than a two-horse team. To be asked by the engineer to fetch something could make a day.

The steam engineer held the place the test pilot or the astronaut does today, but was much more accessible. Auto magnate Henry Ford stated in his 1922 autobiography, "The greatest experience of my young life was encountering a steam engine on the road to Detroit. I was off the wagon and talking to the engineer before my father knew what I was up to. It was that engine that took me into automotive transportation." Ford later became a steam-engine expert, having the job of field service man for G. Westinghouse & Company of Schenectady, New York, and later being the chief engineer for the Edison Illuminating Company in Detroit, Michigan.

Being a steam engineer was a tough, dangerous job. Early engines were made from uncertain metal—metal that could have dangerous weak spots. Construction methods in the early days were also lacking. Rivets could pop, castings could be porous, and the steam vessel might rupture. The engineer was also blamed for horse runaways, bridge collapses, and fires. Mechanical breakdowns required the utmost in ingenuity as parts had to be hand forged in many instances. Dirty water caused problems of foaming, which tended to

1913 Case 110
Below: *"Let 'er smoke!" A 110-hp Case steam engine leans into the traces pulling a twelve-bottom gang plow at the annual Rollag, Minnesota, show.*

carry water into the cylinder. Too much water in the cylinder could result in cylinder breakage. Other impurities in the water coated the heat-transfer surfaces. Suitable fuel was also a continuous problem. Coal was the best, but seldom available. Various kinds of wood did well, but a source close to the job was necessary. Eventually, on the Great Plains, straw was the fuel of choice, but it worked best in engines designed for that fuel.

During their heyday, there were as many as 75,000 custom-thresher engineers. They followed the harvest from south to north. Sacrificing a normal home life, they mostly slept outside under the machines. They returned year after year, lured not by financial reward, but by their love of machinery, the satisfaction of a job well done, the excellent food served by the farm women, and by the high esteem of the farm lads.

Steam power also seemed to have an almost mystical charm for those associated with it—and still does today. When thus infected, a train whistling for a distant grade crossing could not be ignored, nor could the chuckling sound of a steam engine operating a thresher or a sawmill. Old-timers would try to explain their fascination to the younger generation, but explanation was either unnecessary or ineffective: You either had the steam bug, or you were immune.

1918 Sawyer-Massey 76

The Sawyer-Massey Company of Hamilton, Ontario, was one of Canada's leading builders of threshers and steam engines. It all began from a humble blacksmith shop started by John Fisher in 1835. In 1840, a relative, L. D. Sawyer, bought into the fledgling firm, and they began building threshing machines. When Fisher died, the name of the firm was changed to the L. D. Sawyer Company. In 1892, Hart Massey, of Massey-Harris fame, bought a 40 percent interest, and the name was changed to Sawyer-Massey. The Massey family withdrew in 1910, however, but the name continued. Owner: Ontario Ministry of Agriculture and Food.

Wood Brothers

Wood Brothers Inc. of Des Moines, Iowa

1914 Wood Brothers 30

Wood Brothers's 30-hp "double-geared" steamer of 1914 was novel in providing geared engine power to each rear wheel, resulting in steady, balanced pull. Most steam engines of the day were single geared.

F. J. Wood and his brother, R. L., founded the Wood Brothers Thresher Company in Rushford, Minnesota, in 1893. In 1899, they moved their operations to Des Moines, Iowa, where the firm was renamed Wood Brothers Inc. F. J. Wood remained head of the company until he retired in 1945. Shortly thereafter, Henry Ford bought the company.

An inveterate inventor, F. J. Wood conceived of a double-geared, 30-hp traction engine with a center crank, which debuted in 1915. The dual power paths to the drive wheels lessened the strain on each gear. Highly stressed gears were submerged in oil, and other components were designed for durability when plowing.

This 30-hp engine was the largest made by the Wood Brothers. Others followed in the 20- to 25-hp range.

Avery

Avery Company of Peoria, Illinois

Brothers Robert and Cyrus Avery founded the Avery Company in 1874 in Galesburg, Illinois. After ten years in the corn-planter business, they moved their operation to Peoria, Illinois. They built their first steam traction engine in 1891. It was a single-cylinder, straight-flue machine with a robust boiler, allowing higher pressures than the competition. Avery also branched out into the thresher-manufacturing business at that time.

1912 Avery Undermounted 18

Avery's Undermounted was a two-cylinder, locomotive-type engine noted for smoothness and quietness. The firm built engines of this style with up to 50 hp; this engine was rated at 18 hp.

John Bartholomew, a relative of the brothers, became company vice-president in 1893. After the Avery brothers died, Bartholomew, who had been with the company since he was fifteen, became president. He was an energetic man with talents both in mechanical things and in finances. Bartholomew moved Avery into building gasoline tractors, including an unsuccessful attempt at a combination tractor-truck in 1909.

The crowning achievement of Bartholomew's presidency was the Avery Undermounted, which came out in 1903. The Undermounted was a locomotive-style steam traction engine with two cylinders mounted beneath the boiler. Tractors of 16, 18, 20, 30, 40, and 50 hp were built in this arrangement and were noted for their smoothness and quietness. Avery continued building the conventional single-cylinder type as well with power ranging from 12 to 50 hp.

In the early 1920s, Avery fell on hard times, becoming over-extended both in customer credit as well as in its attempts to compete in the gasoline tractor market. It was re-organized as the Avery Power Equipment Company in 1924. Bartholomew died in 1925, but the company did well until 1931 and the depths of the Great Depression. It was again revived and survived until 1941, building the famous Avery Ro-Trak gasoline tractor.

Advance-Rumely

Advance-Rumely Thresher Company of La Porte, Indiana

With the exception of Case, Advance-Rumely built more steam engines than any other firm.

The roots of the Advance-Rumely Thresher Company stretch back to German immigrant Meinrad Rumely's blacksmith shop, founded in La Porte, Indiana, in 1852. Rumely and his brother John began making threshers around 1854. By 1910, the Rumely Company, now in the hands of Dr. Edward Rumely, the grandson of one of its founders, began building the highly successful line of Rumely OilPull kerosene tractors.

Rumely went on to buy out numerous other firms over the years. It acquired the Advance Thresher Company of Battle Creek, Michigan, followed by Gaar, Scott & Co., and the Canadian firm of American-Abell Engine & Thresher Company of Toronto, Ontario. Soon after, the venerable Northwest Thresher Company of Stillwater, Minnesota, was also added. The company had grown too fast, however, and financial difficulties were insurmountable by 1915, when the whole was re-organized as the Advance-Rumely Thresher Company. In 1923, another great name came under the Advance-Rumely banner: the Aultman & Taylor Machinery Company of Mansfield, Ohio. The Advance-Rumely firm was taken over by the Allis-Chalmers Company of Milwaukee, Wisconsin, in 1931.

Rumely built a variety of steam engines over the years, including single- and cross-compound types. Power ranged from 12 to 35 hp under the old rating system, and up to 140 hp in the new system.

Gaar-Scott steamer, 1900s
"Built for the Pioneer Ranch, Macleod, Alberta," read the writing on the water tank of this Gaar-Scott steamer pulling gang plows to break the Canadian prairie. Gaar, Scott & Company of Richmond, Indiana, was founded in 1870 and acquired by Advance-Rumely in 1911. (Glenbow Archives)

1919 Advance-Rumely 22/65
Advance-Rumely built more than 12,000 steam engines. Originally founded in Canada, Advance was taken over by Rumely in 1911. This engine was built at Rumely's works in La Porte, Indiana. Owner: Kurt Umnus.

CHAPTER 2

The Genesis
of the
Gasoline
Tractor,
1889–1920

"The infant gas tractor can stand the emergency endurance test where the horse and the mule fall down. He will pull all your tillage apparatus by moonlight as well as by daylight. If there is no moon all you have to do is to attach a searchlight."
—Barton W. Currie, *The Tractor*, 1916

1916 Avery 12/25
Main photo: *Built by the Avery Company of Peoria, Illinois, from 1912 to 1919, the 12/25 had a transverse two-cylinder, horizontally opposed engine with exhaust draft cooling.*

1919 Avery brochure
Above: *Avery promoted "Motor-Farming" as the way of the future.*

The Age of Invention

1889 Charter

Both photos: *John Charter's Charter Gas Engine Company of Sterling, Illinois, built what is considered the first internal-combustion-engine farm tractor in North America. The 10/20-hp Charter single-cylinder engine used liquid fuel rather than natural gas, as was typical of some earlier "gas" engines. Charter mounted his engine atop a typical Rumely steam-engine chassis. He built six machines in 1889 and shipped them to northwestern United States farmers.*

Timeline

1901: Radio signals span the
Atlantic

1906: San Francisco earthquake
and fire

1906: Alzheimer's disease
identified

1907: Leo Baekeland patents
the first plastic,
known as Bakelite

1908: William Hoover invents
the vacuum cleaner

1909: Radio broadcasts begin

1914: Panama Canal opens

1925: Scotch tape invented

At the dawn of the twentieth century, civilization was enjoying an explosion of technological advances akin to those at the turn of the twenty-first century. In the 1890s, many Americans had fought in the Civil War. They had then seen frontier log cabins and stumpy fields replaced by prosperous farms with frame houses and huge barns. Barely discernible wagon ruts across the prairies were replaced by railroads. In their lifetimes, the telephone, electric light, phonograph, and safety bicycle had been invented. But that was just the beginning. As the internal-combustion engine was developed, the automobile, tractor, and airplane followed.

The horseless carriage, as the automobile was first called in America, got off to a slow start. Automotive terms suggest French primacy—chauffeur, garage, chassis, sedan, and even Detroit, were French terms—and the automobile (another French word) really got going first in France. At the time of the Paris Exposition of 1900, cars were beginning to outnumber horses on Paris's Champs-Elysées. When Henry Ford founded his Ford Motor Company in 1903, the automobile was still considered a plaything for the rich. A good number of common people resented the rich in their noisy, smelly, horse-frightening automobiles. As Woodrow Wilson commented in 1907, "Nothing has spread socialistic feelings in this country more than the automobile, with its picture of arrogance and wealth."

There were two main obstacles to the acceptance of the automobile in the United States. One was price; the other was the lack of decent roads. Ford solved the first problem with his mass-produced Model T. The Good Roads Movement, founded in 1902 by the American Road Builders Association and the American Automobile Association, began to solve the other. What the Good Roads Movement needed was the tractor to power the implements of road building.

The early development of the gasoline tractor followed that of the steam engine. First, stationary engines were mounted to skids to make them portable. Then wheels were added, then a drive mechanism, and then a means of steering. Finally, a drawbar was forged, and the concept was complete.

The first such tractor was built by the Charter Gas Engine Company of Sterling, Illinois. Issued in 1889, John Charter's patent covered the use of liquid fuel, or gasoline. Prior to that, engines used everything from natural gas to coal dust for fuel. The need for lubricants in the machine age had brought forward the petroleum industry. When refining petroleum to make oil and grease, gasoline was a highly volatile byproduct that was distilled off. With the acceptance of gasoline as an engine fuel, both petroleum and engine companies flourished. Charter mounted his gasoline engine on Rumely steam-engine running gear and eventually sold six of these machines.

The first tractor able to propel itself both backwards and forwards was the 1892 Froelich. John Froelich of Froelich, Iowa, mounted a Van Duzen gasoline engine on a Robinson steam-engine frame and devised his own drive and steering systems. Froelich took his machine and his 40-inch (100-cm) Case thresher on a fifty-day threshing run. As he traveled from place to place, he pulled the thresher with the tractor. Once set up, he then powered the thresher by means of a flat belt from the engine's flywheel. In the fifty-day run, Froelich threshed some 72,000 bushels of small grain. Later, Froelich joined with venture capitalists to form the Waterloo Gasoline Traction Engine Company of Waterloo, Iowa. This company later built the famous Waterloo Boy tractor.

1892 Paterson-Case

Left: *Engineer William Paterson of Stockton, California, came to Case to build an experimental gas engine in the early 1890s. The "balanced" engine was a horizontal two-cylinder inline with water-jacketed combustion chambers. The Paterson was one of the first practical gasoline tractors, and it worked—but not well enough for it to enter production. The Paterson suffered from ignition and carburetion problems, the bane of all early gasoline engines. Undeterred, Paterson patented his engine design on October 30, 1894.*

1914 Townsend Model 30/60

The peculiar Townsend oil tractor was built on what was called a "boiler frame." Five different sizes of these steam engine–lookalikes were offered by Roy C. Townsend's Townsend Manufacturing Company of Janesville, Wisconsin. The Model 30/60 featured a two-cylinder, horizontal side-by-side engine that used oil for cooling. The "boiler" contained tubes around which air was circulated to cool the oil. Airflow was induced by the exhaust in the stack.

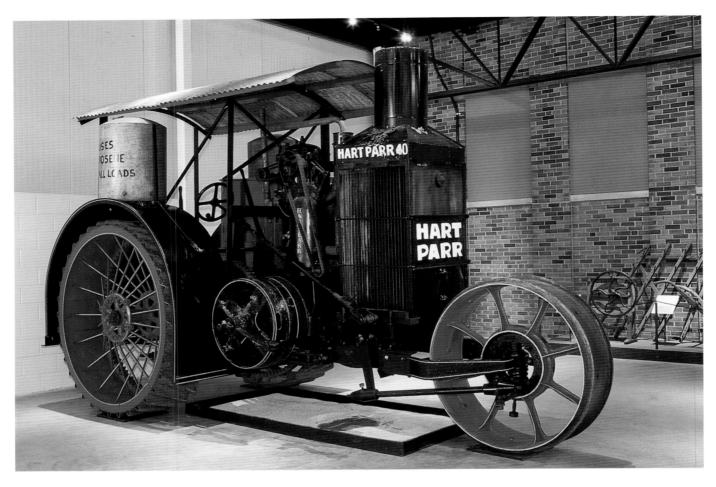

1914 Hart-Parr Model 40 20/40

This Hart-Parr 40 was on display at the Floyd County Historical Society's museum in Charles City, Iowa. This tractor was built in 1914, but others of the type were built from 1912 to 1914. The Model 40 was powered by a vertical, two-cylinder, oil-cooled engine rated at 400 rpm.

1910s Pioneer 30 advertisement

The Pioneer Tractor Manufacturing Company of Winona, Minnesota, began making tractors in 1910 with its Pioneer 30. "First in Gas Traction," promised this ad for the Pioneer's 7x8-inch (175x200-mm) four-cylinder, horizontally opposed engine. The drive wheels on the massive Pioneer 30 were 8 feet (240 cm) in diameter.

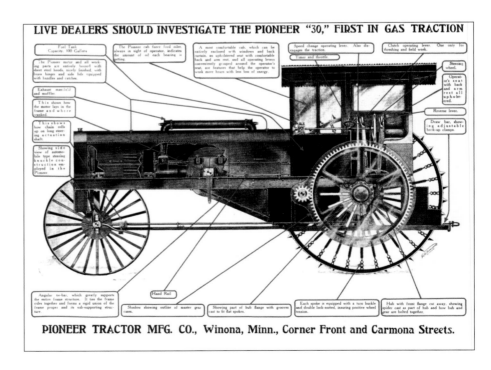

1910s Twin City Model 40

Minneapolis Steel & Machinery Company of Minneapolis was founded in 1902 to make steam engines and other agricultural equipment. MSM entered the gas-tractor business in 1911, naming its machines the Twin City line. Owner: the Vouk family.

Deere & Company of Moline, Illinois, purchased the firm in 1918, thereby jump-starting Deere into the tractor business.

In 1892, Case, well known for steam engines, also made an experimental gas-fueled tractor. This machine was based on an engine design proposed to Case by inventor William Paterson. Paterson had come to Case with a threshing-machine invention. In the course of trying to interest Case in that, he mentioned his idea for a two-cylinder "balanced" engine. Case President Stephen Bull said the firm had no interest in Paterson's threshing ideas, but was interested in the engine.

Case created a chassis design along the lines of its steamers and helped Paterson put his engine together. The Paterson engine had opposed pistons with a crankshaft on one side and linkage from the other side. Despite its complexity, the engine ran smoothly, but ignition and carburetor problems eventually spelled its demise. Case then wisely waited until 1912 to re-enter the gas-tractor business.

In 1897, Charles Hart and Charles Parr founded the Hart-Parr Gasoline Engine Company while they were both engineering students at the University of Wisconsin in Madison. When they had trouble raising capital for expansion, they moved their operations in 1900 to Charles City, Iowa, Hart's home-

1910s Big Four 30 advertisement

The Big Four 30 from the Gas Traction Company of Minneapolis, was one of the most influential of the early gasoline tractors. The gigantic engine featured four cylinders with 4x5-inch (100x125-mm) bore and stroke. Big Four 30 machines were widely sold—especially by Deere & Company branch houses—and even more widely copied by other makers.

town. Since he and his family were well known there, local banks provided the cash necessary for the pair of youngsters to venture into the tractor business.

The duo built their first tractor, Hart-Parr No. 1, in 1901 and sold it in 1902. An improved version, Hart-Parr No. 2, was completed in 1903. Fifteen of the No. 2 were delivered that same year. By 1905, Hart and Parr had established the only business in America devoted exclusively to tractor manufacturing. Hart-Parr Sales Manager W. H. Williams was even credited with coining the word "tractor" in sales brochures in 1906; previously, these devices were called "traction engines." By 1907, one-third of the 600 tractors at work in the United States were Hart-Parrs.

In the first decade of the twentieth century, many of the firms that would have a significant influence on the future of power farming first appeared—including the Wallis Tractor Company of Racine, Wisconsin; Henry Ford & Son of Dearborn, Michigan; McCormick Harvesting Company of Chicago; Wm. Deering & Company of Chicago; Gas Traction Company of Minneapolis, maker of the Big Four tractor; and Minneapolis Steel & Machinery Company of Minneapolis, builder of the Twin City line. Tractors did not have a conventional configuration at this time, so many unusual arrangements of engines and drive systems were on the market. Some of the companies were solidly in the implement business, while others were following a single patent or idea. Ignition systems and carburetion were the big problems; external drive gearing also caused problems. The friction clutch, so common today, defied the ingenuity of designers of the day.

1917 Twin City advertisement
Left, top: *Minneapolis Steel & Machinery offered four sizes in its Twin City line, ranging from 15 to 60 hp.*

1910s Twin City Model 40
Left, both photos: *The Twin City Model 40 was powered by the company's own OHV four-cylinder motor of 1,590 ci (26,044 cc) rated in Nebraska tests at 40 drawbar and 65 belt hp. The tractor weighed more than 25,000 pounds (11,250 kg). The TC-40 was popular with custom thresherman and was built in Minneapolis from 1911 to 1924. The huge radiator on the nose held some 100 gallons (330 liters) of cooling water.*

1916 Russell Giant 30/60

Above: *A Russell Giant pulls a John Deere nine-bottom plow in a plowing contest. The Giant had a vertical four-cylinder engine and two-speed transmission. Russell had been a big player in the steam-engine business before it began making kerosene-burning tractors in 1911 in Massillon, Ohio. The company was founded in 1842 and survived until 1942.*

1913 Avery 40/65

Left: *Built by the Avery Company of Peoria, Illinois, the 40/80 four-cylinder model was a 22,000-pound (9,900-kg) behemoth. The 1,509-ci (24,717-cc) engine proved to be a little short of power on the belt, so the tractor was later rated as a 40/65. It had a production life from 1913 to 1920.*

Besides the technical problems, the automobile and gasoline tractor were causing social upheaval in both cities and rural areas. These were also times of social and political unrest. The Progressive Revolt began in 1896, with both agrarian and urban people fighting against the perceived takeover of the federal government by big business and railroads. The socialists were also gaining political influence among factory workers as Eugene Debs organized his Social Democratic Party in 1896. Most of the problems stemmed from the cities, where slums, crime, and poverty were rapidly spreading and where the exploitation of laborers, especially women and children, was rampant. Immigrants were at the mercy of big business and were also resented by the previous wave of immigrants. Banking empires, mostly unregulated, caused financial fluctuations with little regard for the human suffering they inflicted.

1918 Gray advertisement

1917 Gray 18/36

Above, both photos: *Gray tractors, built by the Gray Tractor Company of Minneapolis, featured a bizarre "wide drive drum" at the rear and a strange sheet metal cover. The drum was 70 inches (175 cm) wide and was designed to distribute the tractor's weight over a broad "footprint" in soft soil, a problem solved by other makers with crawler tracks. The Gray's power came from a 478-ci (7,830-cc) Waukesha four-cylinder motor. The Gray was built between 1914 and 1925. It was an outgrowth of the Knapp Farm Locomotive, which was first built in 1908 in Rochester, New York.*

1920 Hart-Parr 30A 15/30

Hart-Parr built its Type A from 1918 to 1924. It had a two-cylinder, side-by-side horizontal engine and two-speed transmission. The writing on the front of the tractor read "Hart-Parr Co., Founders of the Tractor Industry, Charles City, Iowa U.S.A." Owner: David Preuhs.

1910s Short Turn tractor

Tractors came in all shapes and sizes in the pioneering years of the 1910s and 1920s. The Short Turn 20/30 could turn around in an area its own length at the end of rows. It was the brain-child of inventor John Dahl and was built by the Short Turn Tractor Company in both Bemidji and Minneapolis, in 1916–1918.

1917 Kardell Four-In-One advertisement

The Kardell Tractor & Truck Company of St. Louis, Missouri, offered its innovative Four-In-One machine to serve the lucky owner as a motor plow, truck, tractor, and all-round farm power to operate threshers and other farm machinery. Rated as a 20/32-hp machine, the Four-In-One featured a radical clutch release that operated automatically if the plow hit something solid.

1925 Minneapolis 17/30 Type A

Right: *With its transverse four-cylinder engine, the 17/30 looked much the same as the 27/44, but was some 10 inches (25 cm) shorter and a ton (900 kg) lighter. It was built from 1920 to 1929. Owner: the Timm family.*

1929 Minneapolis 27/44

Below: *The 27/44 boasted two speeds forward, a transverse four-cylinder engine, and a weight of some 4 tons (3,600 kg). Owner: the Timm family.*

Froelich

Waterloo Gasoline Traction Engine Company of Waterloo, Iowa

John Froelich's tractor of 1892 was the first to propel itself both forwards and backwards. Since steam engines were reversible and able to run backwards, a reverse gear and clutch were not required. Simply installing and connecting a gasoline engine on a steamer chassis did not allow for backing up. With something the size and weight of the converted steamers, it is no wonder their market was limited and that the Froelich machine attracted attention.

Froelich used a monster of an engine manufactured by the Van Duzen Gasoline Engine Company of Cincinnati, Ohio, displacing 2,155 ci (35,299 cc) in its single cylinder. The engine was a "square" design with a 14.00x14.00-inch (350x350-mm) bore and stroke. It was mounted vertically to avoid lurching motion when in transit. Ignition came from a contact point and battery system, and it was equipped with a flyball governor that held open the exhaust valve when it ran over the desired speed.

The chassis was built mainly of wood with wheels and gearing from a Robinson steam engine. Froelich designed and built his own clutch and reverse-gear arrangement and steering system. Steering came from the machine's front platform by means of a linkbelt chain used to pivot the front axle. The chain was controlled from a steering wheel through a worm-and-sector gear. Levers were provided at the operator's station to control direction of travel, clutch, and throttle.

The tractor was successful in the hands of Froelich, but after he joined with financial backers to form the Waterloo Gasoline Traction Engine Company and tractors were delivered, many customers were so unhappy they returned the machines for a refund. To generate cash flow, the firm turned to making stationary engines, a venture in which they had a modicum of success. Froelich left the company, however, when the word "Traction" was dropped from the firm's title.

The Waterloo Gasoline Engine Company completed two other tractor designs in 1896 and 1897, but only one of each was built. The firm continued in the engine business until 1912. In 1906, the trademark "Waterloo Boy" was adopted for its engines, and the tractors that emerged from the factory after 1912 were also called Waterloo Boys. It was this company and the Waterloo Boy tractor that was acquired by Deere in 1918.

The Only Gasoline Traction Engine ···ON EARTH.···

The **Waterloo**
⬧ **Gasoline** ⬧
⬧ **Traction** ⬧
⬧ **Engine.** ⬧

·····MANUFACTURED BY·····

The Waterloo Gasoline Traction Engine Co.,
WATERLOO, IOWA, U. S. A.

1890s Froelich brochure
The cover of the original Froelich brochure. Advertisements of the day touted the operator's visibility, since the platform was in front.

1892 Froelich replica
The first practical gas-engined tractor was built in 1892 by John Froelich. Twenty-six years later, it would evolve into the successful Deere tractor line. This replica was constructed by workers at Deere's Waterloo, Iowa, tractor works in 1937 to use in a documentary film. The Froelich replica was on display at Deere's Moline, Illinois, administration center.

1911 Hart-Parr 30/60 "Old Reliable"

Both photos: *The 30/60 was powered by a two-cylinder, horizontal side-by-side engine running at a constant 300 rpm. The "Old Reliable" used a dual jump-spark ignition from a low-tension magneto and a hit-and-miss governor, although the engine was started with dry-cell batteries. There were three tanks built into the 30/60: one for gasoline for starting, one for kerosene (the main fuel), and one for water for water injection. Engine exhaust was discharged into the cooling tower on the front to induce airflow through the radiators and out the chimney on top. This 30/60 is in original, unrestored condition. Owner: Ken Kass.*

Hart-Parr 30/60 "Old Reliable"
Hart-Parr Gasoline Engine Company of Charles City, Iowa

Although Hart-Parr built a full line of oil-cooled tractors between 1901 and 1918, it was the 30/60 "Old Reliable" that truly won the company renown. The 30/60 was introduced in 1907 and was the sixth design to be produced by the young firm.

The engine of the "Old Reliable" was a two-cylinder, horizontal side-by-side, kerosene-burner displacing 2,356 ci (38,591 cc) with a bore and stroke of 10.00x15.00 inches (250x375 mm) and rated at 300 rpm. A 1,000-pound (450-kg) flywheel helped smooth out the uneven firing caused by the pistons going in opposite directions and by the hit-and-miss governor.

Oil was used for cooling as it had a higher boiling temperature than water. Higher temperatures were required for successfully burning kerosene fuels, especially with the jump-spark ignition and low-tension magneto. The cooling-tower radiator was a hallmark of the Hart-Parr oil-cooled tractors for more than fifteen years. Hot oil from the engine was circulated through vertical tubes. Airflow was induced through the tubes and out the top of the tower by the jet-pump effect of the engine exhaust gases being injected in the tower chimney. This method had its roots in steam engines where exhausted steam was released in the smokestack to induce draft in the firebox.

The 30/60 weighed some 20,000 pounds (9,000 kg) and had one forward and one reverse gear. It was a four-wheel machine with swing-axle steering. The differential was in the left final drive.

Hundreds of these tractors were sold, as they were popular on the Great Plains where they were used for pulling the heavy sod-breaker plows. Many also found use in road construction. It was one of the first tractors that could be counted on by engineers with limited knowledge of internal-combustion engines to run day in and day out—hence the moniker "Old Reliable."

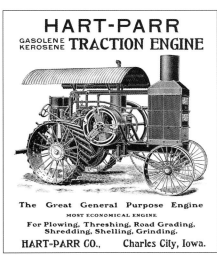

1911 Hart-Parr advertisement
Above: *"The Great General Purpose Engine,"* *noted this Hart-Parr advertisement.*

1913 Hart-Parr 30/60 "Old Reliable"
Above, both photos: *The Hart-Parr 30/60 was manufactured in Charles City, Iowa, from 1907 through 1918. Hart-Parr sales manager W. H. Williams coined the term "tractor" while working on an advertisement for the machine. The 30/60 weighed in at 20,500 pounds (9,225 kg). It had a two-cylinder engine with a bore and stroke of 10x15 inches (250x375 mm) and was cooled by 80 gallons (264 liters) of oil. It gained an enviable reputation for dependability and earned the nickname "Old Reliable." Owner: Gary Spitznogle.*

The Age of Consolidation

The volatility of a pioneering industry such as the fledgling tractor business soon brought on an Age of Consolidation among tractor and farm-equipment makers. An economic recession in the early 1890s prompted whole industries to form trusts among themselves to eliminate competition and mulct the public. The big players in the farm-implement business were squeezed into following suit. In 1891, the Massey Manufacturing Company of Toronto, Ontario, and A. Harris, Son & Company of Brantford, Ontario, were forced to join together as the Massey-Harris Company in Toronto. McCormick, Deering, and five smaller outfits merged in 1902 to form International Harvester Company of Chicago. Case and Deere reorganized, and each acquired smaller companies to keep competitive and avoid being taken over.

Theodore Roosevelt became president of the United States in 1901 after William McKinley was assassinated. He was but forty-three years old at the time, the youngest president up to that point. The first object of Roosevelt's righteous zeal was the "Trust Problem," as it came to be called. Roosevelt invoked the Sherman Anti-Trust Law, which had been on the books since 1890.

International Harvester soon became a target, since five companies from which it was formed had made binders, mowers, and rakes. This, the government maintained, was restraint of trade, and IHC was ordered to dissolve. The company appealed to the U.S. Supreme Court, and legal wrangling dragged on until World War I erupted. The government realized that if it won the case against IHC, the seven other cases pending would automatically follow suit: The result could be the total disruption of war production. Government lawyers were granted an indefinite postponement, but IHC, not wanting this threat hanging over it, sought a settlement. In 1918, IHC agreed to divest itself of the Osborn, Champion, and Milwaukee lines of harvesting machines, and dual McCormick and Deering dealerships were to be eliminated.

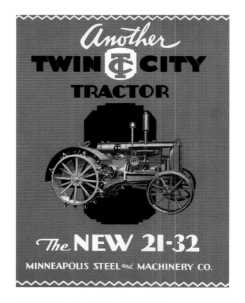

1929 Twin City 21/32
brochure
The venerable Twin City line continued even after the Minneapolis Steel & Machinery Company merged in 1929 with Moline Implement Company of Moline, Illinois, and the Minneapolis Threshing Machine Company of Hopkins, Minnesota, to form the Minneapolis-Moline Power Implement Company.

1918 Twin City 16/30

Left and below: *The TC 16/30 rode on a totally enclosed chassis and a two-speed gearbox. Its L-head engine displaced 589 ci (9,648 cc). A starter and lights were optional. It weighed about 7,800 pounds (3,510 kg). Only 702 16/30s were made between 1918 and 1920. The example shown was built in 1918. Owner: Charles Doty.*

1926 Allis-Chalmers Model E

Allis-Chalmers Company of Milwaukee, Wisconsin, suffered through the post–World War I years and even acquired several other ailing firms. Its Model E tractor featured a four-cylinder, 461-ci (7,551-cc) engine with a 20/35 rating. Owner: Arland Lepper.

1920s Minneapolis 22/44

Above and left: *The Minneapolis 22/44 was introduced in 1921 by the Minneapolis Threshing Machine Company. It had a four-cylinder, 6x7-inch (150x175-mm), OHV engine. It featured an expanding clutch in the belt pulley and a contracting clutch in the flywheel for drawbar work. The 22/44 weighed 12,000 pounds (5,400 kg). Owner: Larry Maasdam.*

1910s Aultman-Taylor brochure

The famed Aultman & Taylor Machinery Company of Mansfield, Ohio, was plagued by financial woes following World War I and was eventually bought out by Advance-Rumely in 1924.

Mogul and Titan
International Harvester Company of Chicago

In 1914, International Harvester brought out its first "lightweight" tractor, the Mogul 8/16, followed in 1915 by the Titan 10/20. At that time, McCormick dealers sold Moguls, while Deering dealers sold Titans. The fact that the 8/16 Mogul was rated for two plows while the 10/20 Titan could handle three led to dealer friction. Therefore, the 1916 Mogul was upgraded to 10/20 hp and a three-plow rating.

The Mogul was a rugged one-cylinder, hopper-cooled workhorse. It featured a channel-iron frame with a kick-up, or arch, in the front. Close-set front wheels were placed under this arch, allowing tight turns. The final drive used a single roller chain with the differential in the left hub. The earlier 8/16 version had only a single-speed transmission with a planetary reverse, but the later 10/20 had two speeds forward with a single reverse.

The Titan used a two-cylinder engine. It had a frame and front wheel arrangement similar to that of the Mogul. Cooling was by the thermosyphon principle, with a large reservoir mounted above the front wheels. A two-speed transmission was used. A double-chain drive ran to the rear wheels.

These so-called lightweight tractors weighed in at around 6,000 pounds (2,700 kg). Nevertheless, they were a lot lighter, cheaper, and more maneuverable than their predecessors. They were also more powerful, affordable, and reliable than most offerings of the time and were an instant hit with the mid-sized farm owner.

1910s International Harvester Titan Model D

Both photos: *The 20-hp Titan D was introduced in 1910 as a single-cylinder machine with a piston displacement of 902 ci (14,775 cc). Weight was 10,000 pounds (4,500 kg). Production ran to 1915, but only 259 were built.*

1908 International Harvester Mogul Type A

All photos: *Built from 1907 to 1911, International Harvester's first gear-drive (as opposed to friction-plate-drive) tractor was known as the Type A. It was built in three sizes: 12-, 15-, and 20-hp.*

1910s International Harvester Mogul 8/16
Sold by McCormick dealers starting in 1914, the Mogul 8/16 used a one-cylinder engine. It featured a chain final drive and single-speed transmission. Later versions were rated at 10/20 hp and had a two-speed transmission.

1919 International Harvester Titan 10/20
The Titan 10/20 was sold by Deering dealers and was a two-cylinder tractor with a two-speed transmission. Like its Mogul sibling, the Titan weighed about 6,000 pounds (2,700 kg).

1920s Mogul and Titan advertisement
International Harvester's English distributor sold both the 20-hp Mogul and Titan whereas in North America, McCormick dealers offered Moguls while Deering dealers had Titans.

**1917 Dain–John Deere
All-Wheel-Drive brochure**

**1916 Dain–John Deere
All-Wheel-Drive prototype**
*By 1916, the Dain–John Deere was in its
third prototype configuration. This one
had the new McVicker-designed engine.
The tractor performed well enough that
production was ordered. The availability
of the Waterloo Boy outfit, the high cost
of the Dain–John Deere, and the death of
Joseph Dain were all factors in the
demise of the tractor.*

Dain All-Wheel-Drive
Deere & Company of Moline, Illinois

Deere's Dain All-Wheel-Drive four-cylinder tractor was conceived in utmost secrecy. The reason for this concealment was the precarious nature of the tractor business in the 1910s as financial backers and bankers considered the business to be high risk.

Further, William Butterworth, Charles Deere's son-in-law who took over the company after the death of Charles, considered himself responsible for the family fortune and was not given to undertaking irresponsible development flyers. But there were powerful voices on the Deere board calling for tractor development. There also had been a successful, if unofficial, relationship between Deere branch houses and the Gas Traction Company in selling its Big 4 tractor as a quasi-Deere product. The result was authorization by Butterworth for low-key tractor experiments.

After an abortive attempt at building a tractor in 1912 designed by board member C. H. Melvin, another board member, Joseph Dain Sr., was asked to create a design in 1915. Dain had previously run his own company selling haying equipment to Deere. The competitive forces brought on by International Harvester prompted Deere to absorb such companies, putting their former owners on Deere's board.

The tractor Dain designed had specific limits put on it by the board. It had to be light, following the lead of the Bull Tractor Company of Minneapolis, whose Little Bull tractor had become the national sales leader. And it was to have a selling price of no more than $700, as the Little Bull sold for $400.

By February 1915, the first prototype was ready for testing. It was an interesting three-wheel design with drive to all three wheels. It weighed a little less than 4,000 pounds (1,800 kg). Originally, a Waukesha motor was employed, along with friction drive. Testing indicated the motor was short of power, and the friction drive would not work.

Second and third variations were built and tested. The friction drive was replaced by an all-gear power-shift transmission, after the fashion of the Model T, except it had two speeds in both forward and reverse. Engineer Walter McVicker was hired to design a new engine, and when this was ready in fall 1916, the tractor had ample power. The new engine had four cylinders with a bore and stroke of 4.50x6.00 inches (112.50x150 mm) and produced 24 belt hp.

Deere's board authorized construction of 100 of the final version of the Dain design, which were built and sold. The future looked bright for the Dain machine, but events quickly conspired against it. Dain died of pneumonia before the 100 tractors were finished, and with him died much of the push for his tractor. The board also became aware of Henry Ford's new low-cost, lightweight Fordson, whereas Deere's Dain would require a $1,600 price tag, far above the $700 goal the board set.

1917 Dain–John Deere All-Wheel-Drive

Besides being an all-wheel drive, the Dain–John Deere featured a two-speed (forward and reverse) power shift. The Dain–John Deere had a lever on the left that was connected to the drawbar. The lever on the right was the gear shift. The first prototypes of the Dain used a Waukesha engine that did not develop enough horsepower. Engine designer Walter McVicker was retained to create the engine eventually used. This is one of only two remaining Dain–John Deere tractors. Approximately 100 of these were built in 1917 and sold in the Huron, South Dakota, area. No record remains as to the disposition of the other 98. This one was donated to the Northern Illinois Steam Power Club of Sycamore, Illinois, by F. L. Williams, now of Sebastopol, California. Club member Bill Karl of Maple Park, Illinois, is the monitor and operator of the machine.

Finally, Deere heard that the Waterloo Gasoline Engine Company, maker of the Waterloo Boy tractor could be bought for $2.3 million. At the time, both the Waterloo Boy Model R and Model N were in production. They had an excellent reputation coupled with a selling price of $985 for the R and $1,150 for the N. On March 14, 1918, Deere bought the Waterloo outfit, rights to the famous Waterloo Boy, and the Waterloo Foundry, a separate but related company.

Waterloo Boy

Waterloo Gasoline Engine Company of Waterloo, Iowa

1917 Waterloo Boy advertisement

Above: *"Farmers are clamoring for the Waterloo Boy," promised this advertisement to potential dealers.*

1910s Waterloo Boy Model R

Facing page: *The Waterloo Boy was a direct descendant of the Froelich tractor of 1892 and was the forefather of the John Deere tractor. It was the first to use the two-cylinder side-by-side engine that would characterize Deere tractors for more than forty years. The Model R Waterloo Boy was built between 1915 and 1919. The Model N was introduced in 1917 with a two-speed transmission and other improvements. Deere & Company bought the Waterloo Boy outfit in 1918. The Waterloo Boy N was continued into 1924, overlapping in production both the Model R and the first two-cylinder Deere, the Model D.*

The well-known Waterloo Boy was a direct descendant of the original Froelich tractor of 1892. The Waterloo Gasoline Traction Engine Company, formed by Froelich and others, did not set the tractor world afire. When the other investors insisted on building engines only, Froelich left the company. Two more abortive attempts at tractors were made, but it was not until 1911 that things began to improve for the new Waterloo Gasoline Engine Company.

In 1911, A. B. Parkhurst of Moline, Illinois, joined the firm, bringing with him some of his own tractor designs. A feature of his creations was a two-cylinder, horizontally opposed, two-cycle engine. Two tractor models were built using this engine, the Models L and LA. The L had three wheels with one-wheel drive; the LA was the same tractor, but converted to four wheels with two-wheel drive.

The two-cycle engine did not prove satisfactory, however, and since the company was familiar with four-cycle engines, it decided to convert the Model LA to a new four-cycle, two-cylinder engine. A horizontally opposed type became much too wide for the chassis, so it was changed to be a side-by-side type. While the former type is naturally balanced and has even firing, the side-by-side type requires compromises. If the two cylinders moved in unison, even firing resulted, but tremendous counterbalances were required. If the cylinders moved in opposition to each other, the counterbalance problem was minimized, but uneven firing resulted. While the competitive International Harvester Titan opted for the cylinders moving in unison, Waterloo chose the other option, casting in iron the characteristic Poppin' Johnny sound of subsequent Waterloo Boy and John Deere two-cylinder tractors. Some believe that this distinctive sound, as much as anything, accounted for their success.

With the new engine, the Model LA became the Model R, built from 1915 to 1919. The engine originally had a bore and stroke of 5.50x7.00 inches (137.50x175 mm), which was changed in 1915 to 6.00x7.00 inches (150x175 mm). In 1917, when the Model N was introduced, the R was switched to the N's larger, 6.50x7.00-inch (162.50x175-mm) engine. The R featured one forward and one reverse speed as well as chain-driven, swing-axle steering.

The Model N was the final version of the "Boy." It could be distinguished from the R by the size of the final drive gear inside of each rear wheel. On the N, which had a two-speed transmission, this gear was as large as possible to maximize torque. On the single-speed R, this gear was much smaller. After 1921, the steering was changed to automobile-type steering. Production continued into 1924.

The name "Waterloo Boy" was believed to have been a play on the then-widely used term "water boy" and the fact that these farm engines were often used to pump water. A lad was employed to run from the pump to the locations where the various men were working, carrying a wooden pail of cold water and a dipper. Needless to say, on a hot, dry harvest day on the Great Plains, the water boy was a welcome sight!

1910s Waterloo Boy Model R

Right: *Kenneth Kass of Dunkerton, Iowa, drives his Waterloo Boy Model R. Ken and his father, Art Kass, farm in east-central Iowa. Art Kass remembers farming with a Waterloo Boy when he was a boy. He noted that the front axle was held on by a pin that allowed for swing-axle steering. Sometimes, the cotter pin on the end of it would wear out—which happened to him on the far end of the farm. Due to a jerky clutch, the front end lurched up and the steering pin fell out, dropping the nose of the tractor on the ground. It was a long walk home to get the team, wagon, and jacks to get it up again. Once he got the tractor back together, he first had to take the team and wagon home, then walk back to the tractor, and resume plowing where he left off.*

1910s Waterloo Boy advertisement

Rumely OilPull

Advance-Rumely Thresher Company of La Porte, Indiana

The OilPull line of tractors began in 1909 when Dr. Edward Rumely, grandson of the company founder, collaborated with John Secor, who had been doing engine experiments for more than twenty years. They developed a line of one- and two-cylinder tractors that used oil, rather than water, as an engine coolant. The oil, with its much higher boiling point, allowed higher engine temperatures for more efficient use of the low-volatility kerosene fuel.

Engine cylinders were offset from the crankshaft centerline to reduce piston side loads. The engines used water injection, sometimes consuming as much water as fuel. Cooling oil was circulated by a centrifugal pump. Engine exhaust was used to induce airflow through the chimney-type radiator. High-tension magneto ignitions were used, except on the early versions. The tractors were driven through an expanding shoe clutch

1928 Rumely OilPull advertisement
The OilPull was "the Symbol of Power and Dependability," according to this Rumely ad.

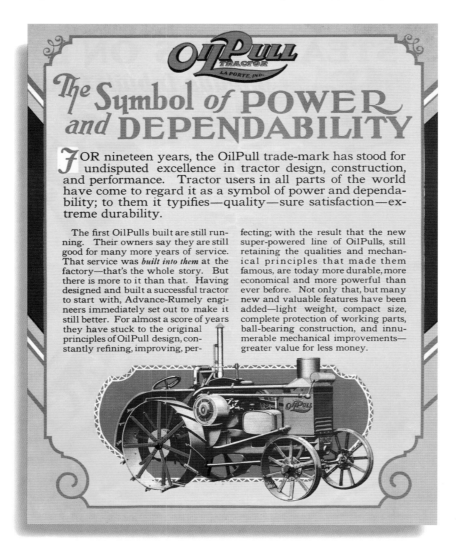

and a spur-gear final drive. One-, two-, and three-speed transmissions were used.

The OilPull line was noted for its large-displacement, slow-turning engines. OilPull tractors could generally best their official power and pull ratings, a fact that endeared them to their owners.

Over the years, Rumely built a long line of OilPull models and variations, including the 25/45 Type B of 1910–1914; one-cylinder 15/30 Type C of 1911–1917; one-cylinder 18/35 Type D built only in 1918; 30/60 Type E of 1911–1923; 14/28 Type F of 1918–1919; 16/30 Type F of 1919–1924; 12/20 Type K of 1919–1924; 20/40 Type G of 1920–1924; 25/45 Type R of 1924–1927; 20/35 Type M of 1924–1927; 15/25 Type L of 1924–1927; 30/60 Type S of 1924–1928; 20/30 Type W of 1928–1930; 25/40 Type X of 1928–1930; 30/50 Type Y of 1928–1930; and 40/60 Type Z of 1929–1930.

1929 Advance-Rumely Type Z
Rated at 40 drawbar and 60 belt hp, the OilPull Type Z used oil for cooling because of its higher boiling temperature. The higher temperatures allowed for better burning of kerosene fuel. Owner: Don Wolf.

1928 *OilPull Magazine*
Left: *Advance-Rumely published an OilPull owner's magazine throughout the 1920s that featured everything from agricultural tips to farming poetry.*

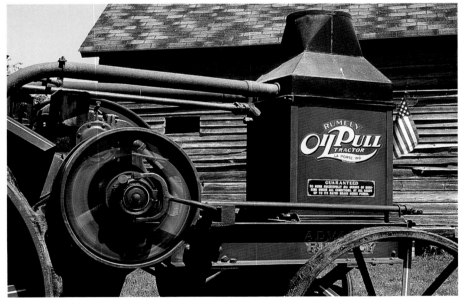

1923 Advance-Rumely OilPull Type G

Above and right: *The 20/40 Type G was introduced in 1920, and production continued through 1924. Like other Rumelys, it used oil as a cooling medium. Engine exhaust induced airflow through the chimney-type radiator. Owner: Glen Braun.*

CHAPTER 3

The **Debut**
of the
Lightweight Tractor, 1913–1935

"I think it safe to eliminate the horse, the mule, the bull team, and the woman, so far as generally furnishing motive power is concerned."
—W. L. Velie, Director of Deere & Company, 1918

1920 Hart-Parr 30A
Main photo: *The 30A of 1920 was a departure from the previous large, oil-cooled tractors from Hart-Parr. It was the first of a line of comparatively lightweight water-cooled machines from the Charles City, Iowa, works.*

1920s Fordson advertisement
Above: *Henry Ford termed his creation the "modern tractor." For its day, the Fordson was indeed revolutionary.*

The Age of the Lightweights

The decade leading up to World War I was one of tentative steps as technology found its footing. Roads, motor cars, tractors, and fuels all were developed together, finally achieving practical efficiency levels. Although pioneered in the United States, the aviation industry flourished mainly in Europe. Electrification came to the city, giving urban people a taste for modern conveniences—and driving a cultural wedge between them and their country cousins.

With the advent of the automobile, country folk were much less isolated, however. The Great War, as it came to be called, had a profound coalescing effect as city and country people came to appreciate their reliance upon one another. After the United States was drawn into the conflagration, the American people realized that they were not as isolated from the rest of the world as had been thought.

In the fifteen years that gas tractors had been available to farmers, two characteristics emerged: First, gas tractors took on the general appearance of steam tractors; and second, like steamers, they got larger and larger. While the Froelich tractor was relatively small—it boasted 20 hp and weighed some 2 tons (1,800 kg)—the average size was soon represented by the monstrous 1913 Avery 40/80, which weighed more than 11 tons (9,900 kg).

The famous Tractor Trials in Winnipeg, Manitoba, focused attention on the big tractors' plowing acreage-per-hour rate. The 21,000-pound (9,450-kg) International Harvester Titan demonstrated a plowing rate of 2.54 acres (1 hectare) per hour in 1913. This, of course, was impressive—especially if you farmed in the Great Plains, or had one of the huge California spreads. But by 1900, fully two-thirds of American farmers had less than 40 acres (16 hectares) of land. By the time the typical farmer pastured and hayed enough land to feed a team of horses or mules and several cows, there was not much land left for cultivation. There was no way most farmers could even consider power farming.

By 1913, several factors combined to change the plight of the small-acreage farmer. The first was the drive to consolidate small farms into larger holdings. The second was the opening of the vast areas of the Great Plains to homesteaders, with 160 acres (64 hectares) of land allotted to each. And finally, in September 1913 a tractor sensation, the Little Bull, was offered for sale at the affordable price of $335. This price was about the same as for a good horse team.

The Bull Tractor Company introduced its lightweight Little Bull at the 1913 Minnesota State Fair. The Little Bull employed a single driving wheel on the right side, an unpowered balancing wheel on the left, and a single front wheel for steering, which was also on the right. The front wheel and drive wheel ran in the previously plowed furrow. A leveling arrangement was included with the balance, or idler wheel, so that the tractor would run level whether or not its two right wheels were in a furrow. A two-cylinder, horizontally opposed engine provided 12 hp. This trim little 3,000-pound (1,350-kg) machine rattled the industry and, by the end of 1914, more than 4,000 had been sold. The Bull Tractor Company was first in sales, displacing the mighty International Harvester.

Prosperity was fleeting for the fledgling Bull firm, however. Most purchasers were howling that they had been stung before the year was out. The

1913 Bull

The Bull Tractor Company of Minneapolis introduced its Bull tractor in 1913, and the lightweight design became influential in the tractor industry for a number of years. The Bull used a large drive "bull" wheel on the right side and an idler wheel on the left, circumventing the need for a differential. The three-wheel design aligned the single front wheel with the right driving wheel. The Bull had a two-cylinder horizontally opposed engine producing 5 drawbar and 12 belt hp.

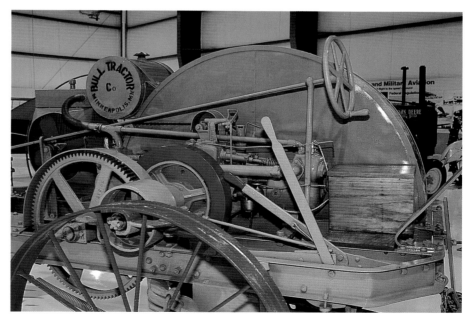

1913 Bull

Above, both photos: *The Bull, or "Little Bull" as it was later known, made waves in the tractor industry because it was light at 3,000 pounds (1,350 kg) and inexpensive at a mere $335. The 5/12-hp tractor had a two-cylinder engine and was of the three-wheel design with single-wheel drive and an idler, or balance, wheel at the rear. The design was updated in 1915 to two-wheel drive with more power and called the Big Bull. Despite the influential design of the Bull, the company did not survive past 1917. This Little Bull is displayed at the Heritage Hall Museum in Owatonna, Minnesota. Owner: Bill Thelemann.*

Little Bull had not been sufficiently tested over a long period: While it performed well enough at the start, failures occurred with time and heavy use. For 1915, the Big Bull was introduced. It had the same configuration, but included a larger, 20-hp engine and beefier components. By 1917, engine power was again increased to 24 hp and the balance wheel was powered. At this point, Massey-Harris became interested in acquiring rights to the Big Bull, but manufacturing difficulties drove the Bull Tractor Company into bankruptcy.

Nevertheless, the Bull's impact on the tractor industry was permanent. A plethora of lightweight machines flooded the market. Numerous machines were unabashed copies of the Little Bull, such as the Model 10/20 from Case; the Little Devil from Hart-Parr; the Model 10/18 from Allis-Chalmers; the Happy Farmer from La Crosse Tractor Company of La Crosse, Wisconsin; the Model L 12/20 from Emerson-Brantingham Implement Company of Rockford, Illinois; and the Light Four from Huber Manufacturing Company of Marion, Ohio. Some were good, some were not. Some were not truly lightweights, such as the 4,000-pound (1,800-kg) Wallis Cub, 5,700-pound (2,565-kg) International Harvester Titan 10/20, and 6,000-pound (2,700-kg) Waterloo Boy. Tractors in this class, although expensive, were more likely to satisfy their owners.

1918 International 8/16
International's lightweight 8/16 looked more like an automobile than a tractor.

One interesting lightweight did enter the field during this period—the Universal tractor. The Universal was introduced in 1913 by the Universal Tractor Company of Columbus, Ohio. Rights to the Universal were quickly bought out in 1915 by the Moline Plow Company of Moline, Illinois, which popularized the machine. Copies of the Universal were also built by Allis-Chalmers as its Model 6/12 and as the Indiana from the Indiana Silo & Tractor Company of Anderson, Indiana.

1919 International 8/16
Built from 1917 through 1922, this tractor featured an inline four-cylinder, 4x5-inch (100x125-mm), OHV engine that was taken from the International Model G truck. In fact the entire configuration was similar to that of the truck with the radiator behind the engine.

1910s Happy Farmer

Above and below left: *The Happy Farmer Tractor Company was incorporated in 1915, with its headquarters in Minneapolis. However, most of the Happy Farmer tractors were made in La Crosse, Wisconsin. It was one of the first of the truly lightweight tractors. Power was provided by a two-cylinder engine of 255 ci (4,177 cc). Owners: Wendell and Charlene Shellabarger.*

1917 Happy Farmer advertisement

1910s Moline Universal Model C

The Moline Plow Company got into the tractor business in 1915 when it bought out the Universal Tractor Company of Columbus, Ohio. Capitalizing on the firm's experiments, Moline brought out a truly unique machine, the Moline Universal. Owner: Walter Keller.

1919 Allis-Chalmers 6/12

Built in 1919, the 6/12 was an unlicensed copy of the Moline Universal—and Moline demanded royalties from Allis-Chalmers. The 6/12 required an implement sulky to hold up the back end. Steering was by an articulation in the middle. Owner: Norm Meinert.

1918 Case 10/20

Like the Bull, Case's tractor had one powered wheel, a front guide wheel on the right, and a "land," or "balance," wheel on the left that could be clutched in for more pull. Case's first four-cylinder, the 10/20 was offered from 1915 to 1922.

1919 Wallis Model K
The Wallis Tractor Company of Racine, Wisconsin, built the Model K, which was sold by the neighboring J. I. Case Plow Works. This Model K was photographed at the Colfax School, part of the Old Midwest Threshers Museum of Mount Pleasant, Iowa. Owner: Alan Buckert.

1910s Hart-Parr Little Devil
The Hart-Parr Little Devil was built between 1914 and 1916, and was powered by a 22-hp, two-cylinder, two-cycle engine and one-wheel drive. It was one of the less successful Hart-Parrs.

1920 Massey-Harris Parrett No. 2

Massey-Harris marketed three versions of the famous Parrett tractor between 1918 and 1923. The Parrett Tractor Company was in Chicago, but castings are stamped "MH," indicating some re-engineering and Canadian manufacture. The tractor had a four-cylinder Buda engine giving it a 12/22 rating. Owners: Rich and Joyce Holicky. Rich bought the tractor in Canada; the former owner had purchased a tract of land and found the Massey No. 2 in the woods.

Universal

Universal Tractor Company of Columbus, Ohio;
Moline Plow Company of Moline, Illinois

1917 Moline Universal advertisement

Moline attempted to sell its Universal by comparing it to farming with a horse team and horsedrawn implements.

Not many Moline Universal tractors ever made their way into the hands of farmers, but the machine nevertheless had a great impact on future tractors. It originated both the all-purpose idea and the concept of integrating the tractor and implement, which is what eventually became the concept behind the ubiquitous three-point hitch.

The nominal Universal Tractor Company developed in 1913 the world's first all-purpose tractor—a tractor that could be employed to do all tasks that could be done by horses on the farm. At the time, other tractors could only pull tillage implements and power other machines via the flat belt and pulley. The Universal could plow, harrow, plant, cultivate, mow, rake, harvest, pull a spreader, and do belt work.

The Universal was a light tractor, weighing less than 3,500 pounds (1,575 kg), and had a unique two-wheel configuration with the engine between the wheels. The driver and implement came behind, the wheels of the implement holding up the back end and providing a place for the driver to ride. Steering was by means of a hinge, or articulation, in the middle and a sector gear driven by the steering wheel.

In 1915, the Moline Plow Company bought the rights to the Universal. As acquired, the tractor needed work. Its two-cylinder, horizontally opposed engine was weak and unreliable. The tractor itself was top heavy and somewhat unstable. Most of all, it was expensive. The Moline firm set about correcting its flaws.

In 1918, Moline launched the Universal Model D with a more-powerful four-cylinder, overhead-valve engine capable of 27 hp—although the tractor was rated at just

1918 Moline Universal Model C

Above and right: *The C version of Moline's Universal was an upgrade from the earlier model. This one had a four-cylinder engine, cement wheel weights, a starter, and lights with an electric governor. It was built from 1918 to 1923. The four-cylinder engine produced over 27 hp at 1,800 rpm. Weight, with the cement, was 3,400 pounds (1,530 kg). Owner: Sue Dougan.*

70

9 drawbar and 18 belt hp. Top-heaviness was corrected by factory-installed cast-concrete wheel weights, and traction was improved through the use of one of the first differential locks. A starter and lights were standard equipment, as was an electric engine governor. Axle clearance for cultivating crops such as corn and cotton was just under 30 inches (75 cm). This high clearance set the Universal apart from other tractors and allowed it to do row-crop work that others could not.

Besides having only a single-speed transmission, the Universal's major shortcoming was its price. The 1920 price tag of $1,325 was three times that of the Fordson. Indeed, it cost more money than the offerings from either Deere or International Harvester, and special implements had to be acquired or made. The Universal's price ultimately caused its demise as it did not survive the great Tractor Price War of the 1920s. The innovative Universal was history by 1923.

1910s Moline Universal Model C

The Moline Universal Model C was powered by a two-cylinder Reliable engine, but it proved somewhat weak, so the subsequent Model D was offered with a four-cylinder motor. Steering was accomplished by means of a center pivot and a sector gear, which caused the tractor-sulky plow combination to bend in the middle.

Doodlebugs, Jitterbugs, and Puddle Jumpers:
Home-Brewed Iron Horses

By 1915, a farm tractor cost anywhere from $400 to more than $4,000. Naturally, quality and reliability went hand in hand with a high price tag. International Harvester's Titan 10/20 cost $900 in 1914; the 12/25 Waterloo Boy Model N cost just under $1,000 in 1916. Both were creditable tractors, but that was a lot of cash for the small farmer.

By 1915, the price of Henry Ford's Model T automobile was down to an amazing $450. It was the most popular car among rural people and had a great emancipating effect on their lifestyle. The Model T was their car of choice partly because of the price, partly because it was rugged and durable and could negotiate the mud and ruts of country roads that left high-priced autos stranded—and partly because Henry Ford was a hero to the common man. Ford knew he was held in high regard by farm folk and used every ploy to further the notion that the Model T was built just for them. One much-publicized photo showed Ford himself buzzing up cordwood using the jacked-up rear wheel of a Model T and a belt to drive a circular saw. This method of powering tools such as choppers and feed grinders was widely touted.

With all this emphasis on the usefulness of the Ford on the farm, it was only natural that wily inventors would take the next step and create adaptations to make the trusty Tin Lizzie do tractor duty. The idea was to provide kits that could quickly convert the family car for pulling implements; then on Saturday afternoon, it could be converted back into a passenger car ready to take the family to Sunday morning church. Such conversions, either factory kits or homemade jobs, came to be known affectionately as "doodlebugs," "jitterbugs," or "puddle jumpers."

Forty-five factory conversion kits were listed in the January 1919 issue of *Chilton's Tractor Index* alone and more were available elsewhere. Of these, none were truly as "convertible" as was hoped. Only a few performed well enough to be considered successful tractors. All suffered from shortcomings inherent in the basic Model T. Simply put, the T was not designed for steady heavy pulling, as was required in plowing. On lighter jobs, such as dragging or raking, it did well. While most of the conversions provided reduction

"Don't let your old Ford or Chevrolet go to waste. Use it to make a practical general-purpose tractor that has the pulling power of from two to four horses, yet costs less than the price of one horse."
—Sears, Roebuck catalog, 1939

1918 Staude Mak-A-Tractor brochure

The E. G. Staude Manufacturing Company of St. Paul, Minnesota, promised that it had signed and notarized affidavits from 363 farmers throughout North America lauding the firm's conversion kit that turned your everyday automobile into an iron-willed farm tractor. Such conversion kits were common in the 1910s and 1920s.

1917 Smith Form-A-Tractor advertisement

All you needed to build your own tractor was the $255 Smith Form-A-Tractor kit—"and a Ford" automobile or truck, as this ad noted in smaller type. The Smith Form-A-Tractor Company was based in Chicago.

gearing to the larger rear wheels, the friction-band planetary transmission was a weak point. Nevertheless, it was the first tractor with power downshift, a forerunner to the Farmall Torque Amplifier of the late 1950s.

With credit often unavailable for the purchase of a real tractor, the farmer was forced into ingenious solutions to his power problems. Homemade doodlebugs were crafted from whatever was available. The vehicle frame was often shortened and a truck rear axle added. The Ford Model A automobile was a popular base, but other brands—and combinations of brands—were used as well. A foreshortened Model A truck could be assembled for as little as $25. Traction would be a problem, but some farmers framed up a box on the back and filled it with cement. Only first gear was useful for work, and because of the brakes, only second was safe on the road. Maneuverability was problematical at best. Adding a second transmission behind the original gave a good range of ratios and provided enough torque to break something downstream, if care was not exercised.

Often, these homemade farm tractors served their ingenious makers for years, if not decades. Some homegrown tractors were mere stopgap vehicles used until the farmer could afford a real, factory-built tractor. Others were well-engineered machines that rose above the sum of their disparate parts to become legends in their farm communities, bestowing renown on the farmer who eschewed the factory-built variety of tractor. Either way, they provided a sign and inspiration to the tractor manufacturers that there was a market for lightweight machines.

Doodlebug

Built about 1936, the "Mogg Tractor" was more refined than the average doodlebug—it had headlights and a front bumper. It was the product of one Peter Mogg, a car dealer from Mt. Pleasant, Michigan. Mogg's son-in-law Harold Taylor is at the controls. (Donald A. Tietz collection)

1910s Doodlebug

This was the fate of most homemade tractors, may they rust in peace. Such doodlebugs were common on smaller farms from the early days of the Model T Ford through the 1950s. Most were made from cut-down trucks. Sometimes two transmissions were used in series to gain enough reduction. Generally, however, first gear was used for just about everything.

1910s New Deal

Above: *The New Deal was basically a factory-made doodlebug, incorporating Model T Ford components. Only about 100 were made by the New Deal Tractor Company of Wyoming, Minnesota, and about six are still in existence. Owner: Dean Zilm.*

1941 Sears, Roebuck Economy Tractor

Left: *As late as 1941, the famous Sears, Roebuck and Co. still offered two "doodlebug" tractor kits through its mail-order catalog. The Economy Tractor was essentially a converted Ford Model A automobile and was powered by the Ford's engine. Sears also offered a "thrifty" conversion kit to build your own farming machine.*

The Decade of the Fordson, 1918–1928

With the advent of World War I in 1914, the armed services acquired all the horses they could find for the war effort; this bolstered the demand for tractors as North American farmers were called upon to provide food for much of the world. Also, many young farm men were inducted into the services, so mechanical help was needed on the farm. Tractor companies flourished.

In 1917, when the United States officially entered the war, tractor production doubled. At the same time, some eighty-five new manufacturers entered the field bringing the total number of U.S. manufacturers to more than two hundred.

Among these myriad makers was one machine that stood out from all the rest—so much so, in fact, that it was perhaps the most significant development in the history of the tractor. It was called the Fordson.

Henry Ford founded his Ford Motor Company in 1903, and in 1908 introduced his revolutionary Model T automobile. Ford then turned his attentions to making tractors.

True to his philosophy that excess weight was the enemy of a motor vehicle, Ford's first tractor, made from car components, weighed less than 2,000 pounds (900 kg), about one-tenth of what others of the time weighed. This tractor, which Ford called the Automobile Plow, proved to be too light for traction and endurance, however. Nevertheless, over the next several years, variations of the Automobile Plow were tested, but no production versions resulted. Following this, several new tractors were tried, which were generally identified by the name of the chief engineer responsible for them, such as Joe Galamb's tractor of 1914 and Eugene Farkas's design of 1916.

The year 1914 was pivotal for the Ford Motor Company and, indeed, the world. Ford introduced assembly-line practices, and the Model T's price came down so it was affordable to the masses. And the year saw the start of World War I in Europe.

Fordson assembly line, 1920s
After production details were settled, the Fordson began rolling off Henry Ford's new assembly lines in numbers never even considered by other tractor makers.

In England, the war sparked an immediate need for farm tractors, and the British Ministry of Munitions (MOM) asked Ford and his tractor-building company, Henry Ford & Son, to send the first 6,000 of his Farkas-designed tractors to Britain. Ford swung into action. A series of pre-production prototypes were built and tested. When by September 1917, tests were still going on, the MOM dispatched Lord Northcliffe to encourage Ford to stop improving and start producing. He was successful, and the first production MOM tractor was delivered on October 8, 1917. By year's end, 254 had been completed. The London *Daily Mail* reported that Ford's tractors were good news for Britain, describing them as "wonderful instruments of war." Although off to a slow start, production of the remainder of the 6,000 MOM tractors (which were not yet identified as Fordsons) were completed in a mere sixty days.

Henry Ford & Son turned to domestic production. Now the name "Fordson" was cast into the radiator head tank. The name, a contraction of Henry Ford & Son for use in transatlantic cable shorthand, had caught the fancy of the senior Ford, and so it became one of the world's most famous trademarks. Another 34,000 Fordsons came off the assembly line in 1918.

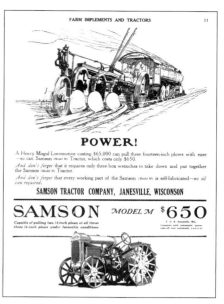

1918 Samson Model M advertisement

Above: *The Fordson became the quintessential tractor of the 1910s and 1920s, and other makers rushed to copy its design. Wanting an entry into the tractor market, Ford arch-rival General Motors of Pontiac, Michigan, bought the Samson Tractor Company of Stockton, California, and moved its operations to Janesville, Wisconsin, where it introduced its Fordson lookalike, the Samson Model M, in 1918. Sadly, the Model M never matched the Fordson's sales record, and GM eventually made a quiet exit from the tractor market.*

1917 Fordson prototype

Above and right: *Henry Ford & Son of Dearborn, Michigan, built sixteen experimental tractors following Ford's acceptance of an order for 6,000 machines from the British Ministry of Munitions. These were built in pairs for testing in late 1917, each pair incorporating improvements found to be necessary in previous testing. These tractors were designed by Eugene Farkas and incorporated a "unit-frame" with the engine, transmission, and rear-end castings being bolted together to form the frame as well as a 251-ci (4,111-cc) Hercules engine and an undershot worm final drive. These machines were not yet Fordsons and had no name emblems applied. The name "Fordson" was applied after the order from the M.O.M. was completed. Owner: Duane Helman.*

Just as he had amazed the world with his revolutionary Model T car, Ford again stunned everyone with the introduction of his radical Fordson. It was conceived as a lightweight, inexpensive tractor to replace not the gigantic steam engines of the large farms but the mules, oxen, and horses of the more numerous small farms. The timing was right, the $785 price was right, and the Fordson became the Model T of farm tractors. By 1920, Ford had 70 percent of the world tractor market.

Fordson

Henry Ford & Son of Dearborn, Michigan

The Fordson was undoubtedly the most significant tractor of the era. With more than 750,000 built in the United States alone from 1917 through 1928, the Fordson was so popular that it virtually personified the normal configuration for tractors of the period. Previously, the configuration was generally that of the Waterloo Boy or IHC Titan.

The Fordson F was the first model built both in the United States and Ireland. It featured a unit frame where the engine, transmission, and axle castings were the frame. It's four-cylinder engine of 251 ci (4,111 cc) produced 10 drawbar and 20 belt hp. It had a three-speed transmission while many of its higher-cost competitors offered only two. It weighed a mere 2,700 pounds (1,215 kg). Over the years, its price would range from a low of $395 to a high of $785; the average price was closer to $500.

After Ford transferred all Fordson production to Cork, Ireland, in 1928, the original Fordson concept was produced in Ireland and England through 1946 with minor modifications as the Model N. The same engine was used in the post–World War II British E27N Fordson Major.

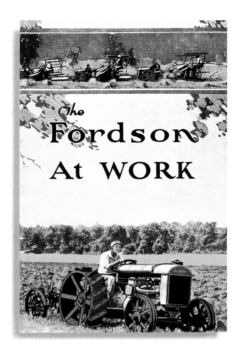

1920s Fordson brochure

There is no question that the Fordson F had shortcomings. The flywheel magneto made them notoriously difficult to start. They had disgraceful traction and the embarrassing tendency to tip over backwards when pulling hard. The use of a worm drive in the rear end caused a howling sound that was recognizable for miles.

Nevertheless, the Fordson introduced most farmers in the United States, Canada, and England to power farming. The Fordson was well loved in Britain where it saved the people from food shortages in two world wars. It was not as appreciated in North America, where farmers readily junked their old Fordsons and switched over to the new row-crop machines of the 1930s.

1918 Fordson Model F

1918 Fordson Model F

Above and left: *The Model F was the first of the Fordsons and the first Ford-built tractor for North American use. To buy one during World War I, a farmer had to obtain a permit from his County War Board. The first of these went to Henry Ford's close friend, botanist Luther Burbank. The Fordson F was much the same as the tractor produced for the British food program, but the engine was soon changed from a Hercules unit to a Ford motor. The earliest Fordsons can be identified by the ladder-side radiator frame. Owner: Duane Helman.*

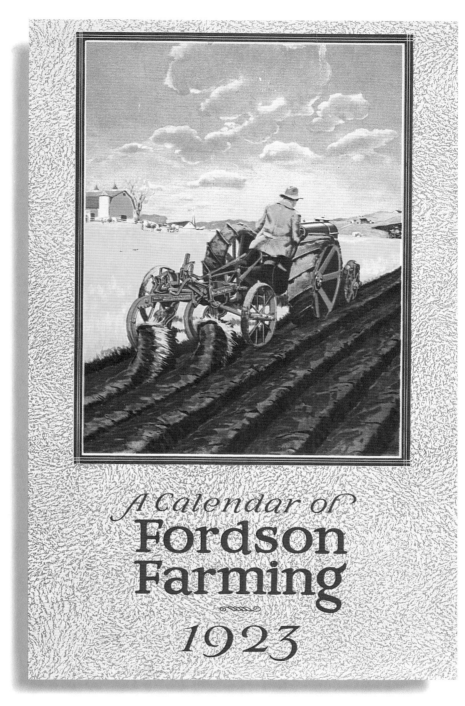

1923 Fordson calendar

THE HENRY FORD STORY: A MIDWESTERN FARM BOY MAKES GOOD

"By the dreams he had, pursued and achieved, the burdens of drudgery were taken from the shoulders of the humble and given to steel and wheel."
—Requiem for Henry Ford by Edgar A. Guest, 1947

If you were born before 1940, you are undoubtedly aware of the aura surrounding Henry Ford. He was reputed to be the richest man in the world. He owned the Ford Motor Company outright. He was famed for being a builder of cars, trucks, airplanes, and tractors. He was respected in political circles; even pressured to run for the U.S. presidency.

If you were born after 1940, you may not have much of an awareness of Ford, the man. You may not know that the Ford Motor Company, founded in 1903, was Henry's third attempt at the automobile business. The first two were failures. (However, his second attempt, the Henry Ford Company, was reorganized to become Cadillac.) You may not know that Henry Leland fired Ford from his second company. Leland later founded Lincoln, and when Lincoln foundered in the early 1920s, Ford bailed Leland out financially—and then promptly fired him.

Such facts and fables of Henry Ford have filled many volumes. But Ford was really two men: He was a cantankerous, self-appointed philosopher; and he was a hands-on, master maker of machines. Like so many other people that rise above their peers, Ford's real genius was his ability to surround himself with loyal followers who had the capability to execute Ford's dreams.

Henry Ford was born in 1863 to Irish immigrants on a prosperous Michigan farm, not far from Detroit. When he died in 1947, his funeral caused a massive traffic jam on the streets of Detroit. These were streets that saw their first car, a Ford, a mere fifty-one years previously. Such were the changes wrought over the eighty-three years of his life. *Time* magazine said, "More than any man in the Twentieth Century, Ford changed the way people lived. He did so by originating a means of getting a useful instrument in many people's hands at lower and lower cost, and in so doing had shown the way to distribute many other useful instruments to millions."

Although Henry Ford's father tried to get him interested in farming, Henry never quite took to it. Instead, he gravitated towards things mechanical, becoming an engineer for the Detroit Edison Illuminating Company. He and his wife, Clara, and their newborn son, Edsel, lived in rented quarters in Detroit. Henry tinkered in his spare time at work, in his shop, and in the kitchen at home. He made himself a two-cylinder engine out of pieces of pipe and other hardware that he found or crafted. With the engine clamped to the kitchen sink, and Clara dribbling gasoline into the intake,

Henry spun the flywheel, and the first Ford engine roared into life just a few feet from the sleeping baby Edsel.

Ford soon had the engine mounted in an automobile, which he called a quadracycle since it looked like two bicycles with a box between. The door to the shop behind the house proved too narrow to get the vehicle out, so an ax and sledge were employed to enlarge the opening of the brick structure. It was near midnight in June 1896 when Henry, followed by a friend on a bicycle, took off on the first test run.

Ford made a variety of cars before hitting on the one that made him famous—the Model T, which made its debut in 1908. At first, the Model T was not all that successful or inexpensive. But in 1914, Ford began using the assembly-line technique, and prices dropped to where it was the most affordable car. He soon had 50 percent of the U.S. market, and over a nineteen-year period, Ford built 15 million Model Ts.

Ford never got along well with the investors in his companies. In 1917, he formed a separate, wholly owned company, Henry Ford & Son, to manufacture his Fordson tractor without stockholder interference. By early 1919, Ford was able to buy out all the non-family stockholders in the Ford Motor Company. He then folded the tractor entity into the car company.

Ford also resisted unionizing his factories. The streets around his Dearborn and River Rouge plants were the scenes of some bloody battles between Ford "guards" and union men. Abruptly, however, in 1941, with his factories ringed by pickets, Ford suddenly acceded to all of the union's demands. Clara, he said, would not let him fight it out. The strike was averted just in time to begin conversion to wartime production.

Ford was an ardent pacifist. Nevertheless, with the bombing of Pearl Harbor, Ford threw himself and his resources into the war effort. With characteristic energy, he turned Ford into an arsenal of democracy. By World War II's end in 1945, his plants and people had built 278,000 jeeps, 8,900 B-24 Liberator bombers, and 57,000 airplane engines.

Before the war ended, however, Ford's only son, Edsel, died suddenly at age fifty. Edsel's son, Henry II, was released from the Navy and called home to help with the business, which because of Henry's crotchetiness and fragility—he was by then over eighty—had always been worrisome to the War Department.

Henry Ford and Automobile Plow, 1907
Henry's 1907 version of the Automobile Plow had an automobile-type radiator, and used the four-cylinder engine and transmission from the luxurious Ford Model B car. The axles and differential came from a Ford Model K six-cylinder car as seen in the background.

After the war, Henry spent most of his time at Greenfield Village, the museum he erected in Dearborn. On April 7, 1947, there was a heavy rainstorm in the area. By evening the basement and power plant of his beloved Fair Lane estate were flooded, leaving the Fords with kerosene lamps and candles for light and fireplace fires for warmth. Ford went to bed early, saying he didn't feel well. He later worsened, called for Clara to bring him a drink of water, and then, just before midnight, he died. With candles for light and wood fires for heat, Henry Ford left the world as he entered it back in 1863, but he had forever changed it nonetheless.

Henry Ford and Fordson, 1910s
Henry Ford, center, was not afraid to get his hands dirty working on the development of his Fordson.

The Minneapolis Ford Tractor and the Nebraska Tractor Tests

1917 Ford advertisement
"At last!! The right tractor at the right price," shouted this ad for the Ford Tractor Company of Minneapolis. The Ford was not the right tractor for Nebraska farmer Wilmot Crozier, however. He was so incensed by the machine's shortcomings that he ran for state legislature and introduced the Nebraska Tractor Test Bill to weed out tractors like the Ford.

Henry Ford tinkered with tractors as soon as he got his Ford Motor Company on solid footing. He boasted to folk that he would make a tractor that would be to the farmer what the Model T was to the public at large. He mentioned such things as a $250 price tag. All of this whetted the appetite of farmers anxious to make the switch to power farming.

Taking advantage of these expectations, a group of investors lead by financier W. Baer Ewing organized the Ford Tractor Company of Minneapolis in 1914. In the organization there really was a man by the name of Ford—Paul B. Ford, to be exact—but he bore no relation to Henry. This Ford knew nothing of tractor design, but was merely recruited so that his name could be used. A few tractors were actually built and sold, but the real purpose of the firm was to force Henry Ford to pay them off so he could use his own name on his forthcoming tractor.

Henry Ford was not about to be trapped in such a scheme, however. Ford and his son, Edsel, launched their tractor enterprise, calling it simply Henry Ford & Son. The Minneapolis Ford outfit soon went into receivership, but not before leaving a legendary legacy in helping to create one of the greatest boons to farmers up until that time—the University of Nebraska Tractor Tests.

Nebraska farmer Wilmot Crozier ordered one of the Minneapolis Ford tractors in 1915. When the tractor was delivered, Crozier had trouble with it almost immediately. Not being one to put up with much, Crozier demanded the company replace it. They did, but the replacement was also unsatisfactory. He then bought a Big Bull tractor. This too was a total disappointment.

Farmer Crozier was beside himself. "I'll try one more before I give up," he said to a neighbor. In 1918, Crozier bought a Rumely "Three-Plow." This tractor was a pleasant surprise, being reliable and durable and regularly pulling five bottoms. The injustice of it all stuck in his craw. Unscrupulous companies were victimizing farmers—who were having enough trouble keeping up, anyway.

Farmer Crozier decided to do something about it. He got himself elected to the Nebraska legislature.

During his first year in office, he teamed with Charles Warner to pass legislation requiring any tractor sold in Nebraska to be tested by the Agricultural Engineering Department of the University of Nebraska. The tests were to ascertain whether or not the machine lived up to advertised claims. If not, such claims could not be made, at least not in Nebraska.

Ag Engineering Department experts L. W. Chase and Claude Shedd devised the tests and test equipment. The first test was attempted in fall 1919, but snowfall prevented completion. The following spring, the Waterloo Boy Model N was the first to be certified. Since that time, the Nebraska Tractor Tests have become world renowned. To this day, tractors by all manufacturers are still submitted to the University of Nebraska for testing.

1910s Waterloo Boy Model R

Above: *Just as the Ford tractor holds an infamous place in tractor history for sparking the Nebraska Tractor Tests, the Waterloo Boy wears laurels as the first tractor to be tested. A Waterloo Boy Model N 12/25 was tested from March 31 to April 9, 1920, and passed with flying colors.*

Wilmot F. Crozier

Left: *Farmer-turned-legislator Crozier was a key figure in the history of truth in advertising for tractors.*

The Great 1920s Tractor Price War

"What? What's that? How much? Two hundred and thirty dollars? Well, I'll be . . . What'll we do about it? Do? Why damn it all, meet him, of course! We're going to stay in the tractor business. Yes, cut $230. Both models. Yes, both. And, say, listen, make it good! We'll throw in a plow as well."

This quotation is from Cyrus McCormick III's great 1931 book, *The Century of the Reaper,* in which he recounts one half of a telephone conversation between International Harvester's Springfield, Ohio, office and the Chicago headquarters. The words are those of the company's hard-boiled general manager, Alexander Legge, who was visiting the Springfield works. This January 28, 1922, telephone conversation came on the eve of the 1922 National Tractor Show in Minneapolis where Henry Ford had just announced a price cut of $230, bringing the cost of a Fordson to the unheard of amount of $395. Most tractor companies spent more than that on their engines.

McCormick went on, "Harvester was waging the battle of the implement industry against mighty Henry Ford and the automobile. Ford was backed by the most popular commercial name of the time and the uncounted millions earned for him by his epoch-making car . . . ."

Harvester did fight back by cutting prices on its popular Titan and International 8/16 by $230 each. They still cost almost twice as much as the Fordson, but the inclusion of the P&O plow sweetened the deal. Harvester's sales people made every effort to turn Fordson demonstrations into field competitions. Harvester and other companies intent on survival, adopted Ford's manufacturing practices.

The real winner of the war was the farmer. The intense competition not only drove prices down, but led to great improvements in parts interchangeability, dealer responsiveness, and in tractor performance. Weaker companies were eliminated in this "survival-of-the-fittest" atmosphere. By 1929, only forty-seven of the original two hundred tractor builders were left.

The cause of the 1920s Tractor Price War was a post–World War I depression that started in 1921. The downturn cut tractor sales in half from their peak in 1920. At 6,000 tractors per month, Henry Ford & Son's shipping lots were jammed by the time the existence of the recession was realized. Ford had no alternative but to cut prices.

This was also a problem for the folks at Deere & Company, who had taken on the Waterloo Boy in 1918. Deere planned to sell forty tractors daily for the last half of 1921. In the end, however, it only sold seventy-nine tractors for the whole model year.

For Ford, the 1922 price cuts that signaled the start of the Tractor Price War were a success. At $395, the Fordson sold. In fact, more than 100,000 were sold in 1923 alone.

The great Tractor Price War was over by 1929. As the recession diminished, retail prices rose swiftly to profit-making levels. By 1929, annual domestic tractor sales were again around 200,000 units.

The Fordson, however, was not among them, as all production was transferred to Ford's new Cork, Ireland, plant in 1928. The Fordson, like the Model T car, had succumbed to determined competition. The new Ford Model A replaced the venerable Model T. The new "general-purpose" tractors—like International Harvester's Farmall—were what farmers wanted now.

1919 Case 10/18 Crossmotor
Case introduced its novel and innovative Crossmotor line in 1915 and soon reaped the rewards of the new machines. The 10/18 was a successor to Case's short-lived Model 9/18B and was built from 1917 to 1920. The two were much the same, except for engine improvements in the 10/18. Both used Case's rock-solid cast-iron frame and novel crossmotor engine installation.

1924 Fordson Model F
The Fordson tractor was essentially mature by 1924. Appearance was somewhat different from previous-year models in that a lighter shade of gray paint was used, and fenders became an option. Wheels were bright red, and the rears had seven spokes. A two-bung fuel tank was used beginning in 1924, with the small gasoline tank inside the main kerosene tank. A hard-rubber steering wheel rim replaced the wood. As before, the Fordson had a 251-ci (4,111-cc) four-cylinder, L-head engine; a three-speed transmission; and the infamous, noisy, and troublesome worm-gear final drive. The worm final drive was retained on the British Fordsons up to 1946. American production of the Fordson ran from 1918 to 1926. Owner: Palmer Fossum.

Orphan Tractors: The First Industry Debacle

1903 Flour City advertisement

Named for its home city of Minneapolis—the flour-milling capital of the world at the time—the Flour City tractor was built by the Kinnard-Haines Company of Minneapolis. The firm began building gas engines as early as 1896 but was forced from the field during the Great Depression.

In the 1910s, North America boasted of some two hundred tractor makers. By 1930, that number had been slashed to about forty-seven. Tough economic times in the years following World War I toppled many pioneer tractor manufacturers; the Great Depression of the 1930s wiped out numerous others who were still standing.

Throughout the ongoing history of the farm tractor, there has been a continuous trimming of the tractor's family tree. Many of those firms that fell out along the way built credible tractors; other machines, however, were bizarre Rube Goldberg creations, imperfect designs, or the brainchild of an engineer who marched to a different drummer. Many fledgling companies over-extended themselves financially as they sought to market their machines; others were simply victims of the numerous economic recessions. Despite financially strong parents, further tractor firms were victims of the intense competition in a marketplace where only the fittest survived.

These orphans—tractors not absorbed or acquired by surviving companies—are today some of the machines that collectors avidly seek out. Yet the orphan tractors are for the truly serious collector, as no help with parts or data can be expected. The reward in satisfaction and acclaim is proportionally great, however, as the restorer may own an incredibly rare or historically important machine.

1917 Interstate Plow Boy and Plow Man advertisement

Uncle Sam himself promised dealers that "America's Demand for Power Farming is Your Opportunity!" Many pioneering tractor makers in the 1910s heeded these words, but they were wiped out by the economic depression following World War I—including the Interstate Engine & Tractor Company of Waterloo, Iowa, that produced the Plow Boy 10/20 and Plow Man 13/30.

1919 Avery Model C

Above: *The Avery Company of Peoria, Illinois, had tunnel vision during the 1910s and 1920s: While other makers were experimenting with innovative lightweights, Avery stayed true to its outdated heavyweights. The firm finally introduced this six-cylinder cultivating tractor in 1919, but it was too late. Like many other smaller companies, Avery filed for bankruptcy shortly after World War I, re-organized in 1925, then closed its doors forever during the Great Depression of the 1930s.*

1919 Nilson 22/45

Left: *The unique Nilson 22/45 was made in Minneapolis. It featured a four-cylinder engine, two-speed gearbox, and three rear driving wheels. This Nilson was on display at the LeSueur, Minnesota, Pioneer Power Show in 1999.*

1921 Huber Super Four

Facing page: *The Huber Manufacturing Company of Marion, Ohio, made a number of tractors of various configurations called the Super Four. This example was the first version. It used a transverse-mounted Midwest engine, which gave it a 15/30 rating. This configuration was produced through 1924, then engine improvements resulted in a rating of 18/36. In 1926, a Super Four 18/36 was introduced with the engine installed parallel to the tractor's centerline. During the Great Depression of the 1930s, however, Huber made its exit from the tractor field to concentrate on construction equipment. Owner: Cliff Peterson.*

1919 Wheat advertisement

Above: *Numerous firms rushed into the booming tractor market in the 1910s and 1920s, but the economic downturn following World War I weeded out the serious makers from the chaff. The Wheat tractor had a short production life in the early 1920s.*

1920s Little Giant

Left: *The Little Giant tractor from the Little Giant Company of Mankato, Minnesota, debuted in 1918 and won a strong following throughout the 1920s. Faced with the Tractor Price War of the 1920s and the roller-coaster ride that the economy took throughout the decade, the company ended production in 1927.*

McCormick-Deering 15/30

International Harvester Company of Chicago

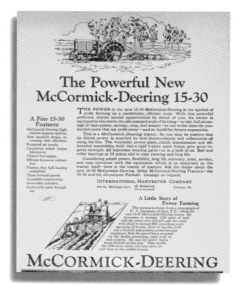

1929 McCormick-Deering 15/30 advertisement

McCormick-Deering 15/30, 1920s

A McCormick-Deering 15/30 pulls a combined harvester-thresher through a wheatfield on the Great Plains.

The big 15/30 was introduced in 1921 to replace International Harvester's Mogul and Titan. It was designed to provide a modern alternative to the Fordson, which was ominously eroding Harvester's market share. While the 15/30 was expensive at about $900, it was everything the Fordson was not. On farms of more than 100 acres (40 hectares), the 15/30 could replace four teams of horses. In the early 1920s, a team cost about $200, so the price was really not insurmountable.

The 15/30 set the industry standard for reliability and durability. Its four-cylinder, overhead-valve, 284-ci (4,652-cc) engine ran on ball-bearing mains. Like the Fordson, it featured a unit frame, but its seat and steering wheel were offset to the right to enhance visibility when plowing.

The McCormick-Deering 15/30 weighed in at 6,000 pounds (2,700 kg), more than twice a Fordson. On the other hand, it was well balanced and boasted excellent traction, which the Fordson did not. It was equipped with a rear power takeoff (PTO) to power all kinds of trailing harvester implements.

Production continued through 1928 when an uprated version, the 22/36 was introduced with engine improvements. The 22/36 was built through 1934.

1920s McCormick-Deering 10/20

Right and below: *Built from 1923 to 1939, the 10/20 was McCormick-Deering's answer to the onslaught of the Fordson. It was a smaller version of the 15/30—14 inches (35 cm) shorter and a ton (900 kg) lighter. Its 284-ci (4,652-cc) engine was slightly larger than the Fordson's 251-ci (4,111-cc). Owner: Fred McBride.*

John Deere Model D

Deere & Company of Moline, Illinois

The John Deere Model D is famous for having the longest production run of any tractor. It was built from 1923 through 1953—thirty years. Certainly, the tractor was much improved over that time, but the original concept was still in evidence at the end of production.

Like many of the tractors of the era, the Model D was designed to overcome the Fordson's onslaught. The D featured a carlike hood and radiator, but under the skin, it was much like the Waterloo Boy it replaced. The Deere was unique among its competitors in that it was powered by a two-cylinder engine. The engine originally displaced 465 ci (7,617 cc); this was increased to 501 ci (8,206 cc) in 1927. It used the Waterloo Boy's two-speed transmission

up to 1935 when it was replaced by a three-speed unit. Following International Harvester's example, the steering wheel and seat were moved in 1931 from the left to the right to enhance plowing visibility. During its first two years of production, the exposed external flywheel had spokes; solid flywheels were used thereafter. In 1939, styled sheet metal was added.

The big, slow-turning engine in the Model D made a delightful sound when plowing. It ran at a mere 800 rpm; in 1932, engine speed was increased to 900 rpm. Besides the lower costs inherent to the two-cylinder engine, the exhaust note was a big factor in its popularity.

The D, like all of the Deere line, was exceptionally rugged. It was designed from the outset to pull three- or four-bottom plows. At the outset, the D weighed 4,000 pounds (1,800 kg)—a ton (900 kg) less than the McCormick-Deering 15/30 although they were in the same power class. As time went on, the weight grew to 5,300 pounds (2,385 kg) without ballast, still far less than the 15/30.

Above: 1939 John Deere Model D brochure

1924 John Deere Model D
Left: *After many false starts and Deere's successful Waterloo Boy acquisition, the Model D became the first truly successful Deere tractor. The Model D introduced in 1923 had a 26-inch (65-cm) spoked flywheel and ladder-side radiator; this successor had left-hand steering and a 24-inch (60-cm) spoked flywheel. The Model D was produced with several improvements for the next thirty years. Owner: Deere & Company.*

Above: 1927 John Deere Model D brochure

1920s John Deere Model D flyer
Left: *Miniature farmers swarm over a giant Model D in this advertisement reminiscent of* Gulliver's Travels.

1932 John Deere Model GP

Above: *The Model GP tractor was made between 1928 and 1935, and was Deere's first answer to the Farmall. This GP, serial number 228863, was one of only 385 Deere GP tractors made in 1932. It is carrying a GP301 check-row planter, made between 1928 and 1935. Owner: the Keller family.*

Right: 1930 John Deere Model GP brochure

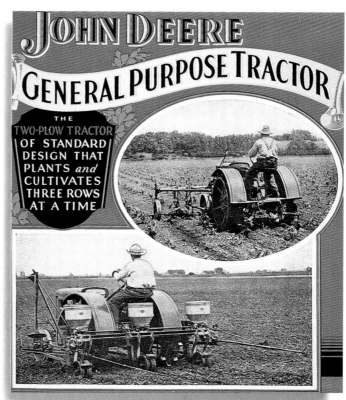

Farewell to the Horse, Hello to the Farmall

It was a sad time for "Old Dobbin": The general-purpose tractor was rapidly becoming the horse's replacement. Once farmers realized that horses were not needed on the farm, a great change took place. Acreage was added to the farm whenever possible, since more could be done with the same manpower. Farmers with extra tractor- and manpower did custom work. While farmers were, in some cases, reluctant to give up their horses, which had struggled along with them through good times and bad, the economics of it became too obvious to ignore.

This change was brought about in large part by the tractor makers. With the advent of the lightweight tractor, manufacturers recognized that their competition was not with the larger tractors, but with the horse. For every 15 acres (6 hectares) a farmer had under cultivation, one horse was needed. Hence, tractor power and indeed price was based on animal competition. The more horses a tractor could replace, the more dollars it could command. The advent of the Fordson and the Tractor Price War provided the impetus for International Harvester to come out with the first truly practical general-purpose tractor—the Farmall.

When the Fordson threatened the whole International Harvester empire in 1922, General Manager Alexander Legge called in Experimental Department Chief Engineer Edward A. Johnston. "What has happened," he asked, "to those versatile tractor concepts you have been working on for the last ten years?"

Johnston and his team had been experimenting with various designs of motor cultivators, some of which were quite innovative. He had, of course, kept Legge and other officials informed of his activities, but there was little corporate interest. In about 1920, the best features of these were combined into an all-purpose tractor, which the team called the Farmall. Several prototypes were constructed. When Johnston told this to Legge and assured him that the Farmall could best the Fordson in every way, Legge ordered twenty more examples built for testing, along with a full complement of implements. The engineers quickly worked on the design to improve performance as well as to make it compatible with new mass-production manufacturing techniques.

The Farmall, as it emerged from Johnston's shop, was built like a tall tricycle. It had large steel rear wheels. The rear axle did not run between the hubs like conventional tractors, but enclosed gear meshes at each end that drove the wheels. Thus, the rear axle could be up and out of the corn. The two front wheels were close together to run between two rows. The tricycle front wheels could rotate almost 90 degrees for steering, making the Farmall much more maneuverable than conventional tractors. In addition, turning the front wheels to their limits actuated cable-operated steering brakes.

Although the Farmall appeared spindly, it was heavier than it looked. It weighed 3,650 pounds (1,643 kg)—about 1,000 pounds (450 kg) more than a Fordson—but the weight was in the right place for traction and stability, two of the Fordson's most glaring shortcomings. Most importantly, the Farmall, as Cyrus McCormick III said in his book, *The Century of the Reaper*, "could handle all farm tasks except milking."

In 1923, more prototypes were ordered, and in 1924, a run of 200 pre-production models were built and sold to farmers. IHC representatives watched closely as the farmers put the Farmall to work, and changes were made as problems arose. The field representatives sent back estimates that the Farmall could reduce costs sixfold when compared to horse farming.

Farmall sales exceeded expectations in 1924, and by 1926, IHC's new Rock Island, Illinois, plant was on line, and Farmalls were rolling out the door. Sales averaged about 24,000 units per year, a far cry from the 100,000-plus Fordsons built annually during its heyday, but along with International Harvester's other tractors it was enough to put it ahead of its nearest competitor, John Deere, by a factor of three.

Junior and Farmall
After old Dobbin made way for the Farmall down on the farm, Junior had a new best friend. (Library of Congress)

Farmall
International Harvester Company of Chicago

1936 Farmall F-12
advertisement

1942 Farmall advertisement

The Farmall heralded the general-purpose tractor revolution when it was first introduced in 1924. It was well thought out and tested before large numbers got into farmers' hands. Thus, Farmalls were noted for living up to their owners' expectations, which more than anything gave farmers faith in power farming and spurred them to retire their horses.

Although not rated by the factory, except for its two 14-inch (35-cm) bottom plow capability, the original Farmall's four-cylinder, overhead-valve engine developed about 25 hp. It weighed a little less than 4,000 pounds (1,800 kg) and was a well-balanced machine. A four-speed transmission was offered. An engine-driven PTO was standard, as was a belt pulley.

Farmalls continued to set the pace for the industry as long as row-crop tractors were popular. Over the years, Farmalls were made in small, medium, and large sizes as well as high-crop versions with even more crop clearance. They became so popular that standard-tread versions of the three sizes were also made available, as were orchard and industrial versions.

Debuting in 1939, the Raymond Loewy–styled Farmalls were so strikingly beautiful they still look modern today.

1930 Farmall Regular

The Farmall was the brainchild of International Harvester's Experimental Department Chief Edward Johnston. It was developed to replace not just some horses on a farm, but all of the horses on the farm. It was introduced at a time when the inexpensive Fordson was glutting the market. Like many other tractor makers, International Harvester needed something dramatic for a counter punch. The Farmall was just that, changing the conventional configuration of the farm tractor from the standard tread to the tricycle row-crop. Owner: Larry Kinsey.

The first of the famous Farmall line was eventually called the "Regular" when subsequent versions were given identifiers. The Regular was built from 1924 to 1931, when it was replaced by the Farmall F-20. Although not rated by the manufacturer, the four-cylinder OHV engine gave the Regular about 20 hp on the belt. Owner: Larry Kinsey.

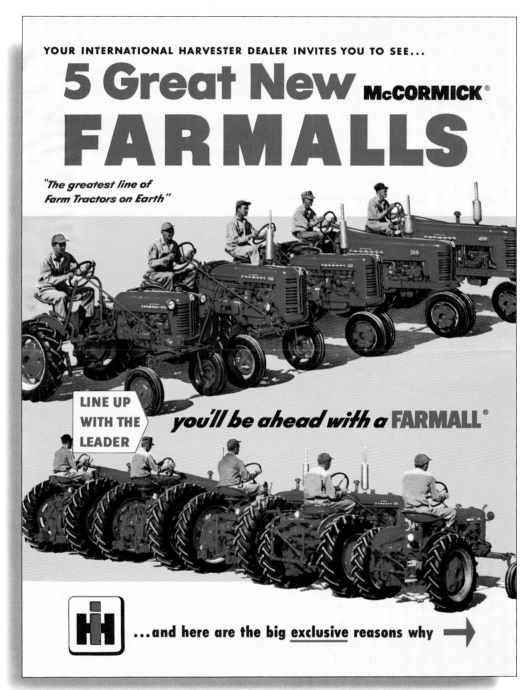

1950s Farmall brochure
"The greatest line of Farm Tractors on Earth," promised this International Harvester brochure. IHC was now offering five Farmall models: the Cub, 100, 200, 300, and 400.

Those New-Fangled Rubber Tires

1934 Allis-Chalmers Model U
*Owner Jim Polacek pulls his 1948 Belle
City Perfection No. 1 thresher to the next
setup.*

1930s Model U tractor race
*Allis-Chalmers organized tractor races at
county and state fairs across North
America to display the speeds possible
with the new rubber-tired Model U.
Here, famed race-car driver and Allis
promoter Barney Oldfield leads two
other Us around a horse-racing dirt-
track. Oldfield always won his race—it
was written into his contract.*

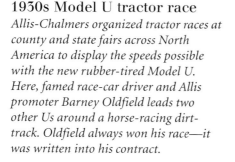

The "Farmer's Friend"—the Model T Ford automobile—rode on rubber tires, prompting the thought, why not use rubber tires on tractors? In the Model T's early days, rubber tires were not without their problems. People carried as many as four spare tires on trips of any distance, along with a complete tire-repair kit. Nevertheless, the idea persisted. Doodlebug tractors used rubber front wheels, and homemade doodlebugs sometimes used truck rear tires.

Non-pneumatic rubber tires (solid bands of rubber) were also used on some tractors. Early Fordsons and other standard tractors were often used as industrial "mules" on solid rubber tires. As early as 1871, a Thompson Steamer was entered in the California State Fair plowing contest using rubber blocks like cleats around both the front and rear wheels. Hard rubber did not provide the needed traction in soft earth, however.

The next avenue to be explored was prompted by orchard growers. Steel lugs were damaging tree roots during cultivation, so orchard tenders mounted discarded truck tire casings on their steel wheels so the natural strength of the rubber arch supported the weight of the tractor; several casings could be used side by side to get enough support. These proved so successful that in 1931 tire maker B. F. Goodrich brought out what was called a "zero-pressure tire."

As early as 1930, Allis-Chalmers began experimenting with real pneumatic tires. A-C engineers mounted a set of Firestone airplane tires on the rear of an Allis-Chalmers Model U. By working with tire manufacturer Firestone, these were developed into true low-pressure tractor tires. Allis-Chalmers then announced that for the 1931 model year, rubber tires would be standard equipment on the Model U.

Nevertheless, farmers were reluctant to make the change—some even offered dire predictions of polluted land unable to grow crops. Of course, some had made the same prediction about the tractor itself.

By 1934, all major manufacturers offered optional rubber tires. By 1936, 31 percent of all tractors were delivered on rubber, and by 1939 almost all were rubber equipped. During World War II, the rubber shortage forced many farmers to go back to steel, but as soon as possible, they returned to rubber.

1936 Allis-Chalmers Model U

The Model U was originally designed and built in 1929 by Allis-Chalmers for the United Tractor and Equipment Corporation of Chicago, but A-C took over the U completely when United folded. At the same time, Allis worked with Firestone in developing pneumatic tires, and the Model U became the first tractor to be equipped with the new-fangled rubber. Up to 1932, a flat-head Continental engine was used in the Model U. After that an OHV Allis-Chalmers engine was used. Production was continued through 1944. Owner: Alan Draper.

1938 Case CC-3
The CC was Case's first true general-purpose tractor. Owner: Jay Foxworthy.

1938 Minneapolis-Moline Twin City FT-A

Left: *"A tractor is like a horse—it pays to buy a good one,"* was the M-M advertising slogan in the 1930s. The FT-A was, indeed, a good one. It was introduced in 1929 by the Minneapolis Steel and Machinery Company, which called its line of tractors the Twin City. Originally known as the Twin City 21/32, the FT-A featured hardened steel gears in its three-speed transmission, dual air cleaners, and full-pressure lubrication. The designation was changed to FT-A in 1936 to be more in line with other M-M designations. Owner: the Timm family.

1937 Fordson Model N

Above: *Henry Ford moved Fordson production to Cork, Ireland, in 1928 and, soon after, to Dagenham, England. Fordsons were then painted blue with orange trim. The Model N's four-cylinder flathead engine displaced 267 ci (4,373 cc). Owner: Eric Coates.*

1938 Fordson All-Around

Left: *Fordson offered its All-Around row-crop variation to satisfy demand in the North American market. Most All-Arounds were exported from England to the United States. Owner: Marlo Remme.*

CHAPTER 4

Farm Crawlers, 1900–1960

"She crawls along like a caterpillar."
—Charles Clements, 1905

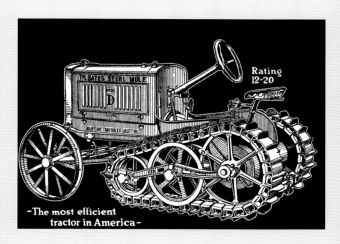

Rating 12-20

—The most efficient tractor in America—

1937 Caterpillar Twenty-Two

Main photo: *Call it a tracklayer, crawler, or caterpillar, the steel-tracked tractor was created to spread the machine's weight over a larger surface, increase traction, and lessen soil compaction. Caterpillar's Twenty-Two had a 251-ci (4,111-cc), four-cylinder engine with overhead valves. It was virtually the same tractor as the R2, built in limited numbers for the U.S. government. Owner: Marvin Fery. Fery was one of the founders of the Antique Caterpillar Machinery Owners Club.*

1920s Bates Steel Mule 12/20 advertisement

Above: *Made by the Joliet Oil Tractor Company of Joliet, Illinois, the Steel Mule may have lived up to its namesake's proverbial stubbornness as it soon disappeared from the market.*

Caterpillars, Crawlers, and Tracklayers

Charles Clements, a photographer hired by Benjamin Holt to record scenes of his new crawler tractor for a brochure, coined the term "caterpillar" for the crawler tractor. Clements's view through his primitive camera was of an upside-down machine, the undulation of the topside of the track links reinforcing the "caterpillar" notion. Holt seized upon the descriptive term immediately and made it the now-famous trademark.

By 1905, when Holt made his first steam Caterpillar, crawlers were not a new idea. Attempts had begun in the 1850s to enlarge the footprints of the heavy steam traction engines, many of which tipped the scale at upwards of 50,000 pounds (22,500 kg). The first appears to have been the Minnis Steam traction engine of 1869, which was demonstrated in Ames, Iowa, but nothing more was heard of it.

There were other attempts after the Minnis, but the first to gain success was Alvin Oliver Lombard's Lombard Log Hauler of 1900 from the Waterville Iron Works of Penobscot County, Maine. It was a half-track steam engine with skis in front for steering. One of the reasons for its success was that the machine was only driven on snow and ice, therefore there was little grit to wear out the track pins and little resistance to sliding the tracks sideways for turning. It is interesting to note that the "pilot" sat in front where the cowcatcher would be on a railroad locomotive, controlling the skis with a steering wheel. The engineer rode in the cab, handling the throttle and reversing (brakes) control. About 200 Lombards were sold, and their patents were licensed to others who made similar machines.

Timeline

1903: Wright brothers fly their first airplane

1928: Walt Disney Productions introduces Mickey Mouse

1937: Amelia Earhart lost over the Pacific

1939: First flight by a jet plane, built by Germany's Heinkel Aircraft Company

1941: Aerosol spray can invented

1948: Transistor invented

1956: Interstate highway system approved

1957: Soviets launch first satellite

1960: First laser operated by U.S. physicist Theodore Maiman

1922 Fordson with Hadfield-Penfield crawler conversion
The lightweight Fordson was notorious for the lack of traction. The Hadfield-Penfield Steel Company of Bucyrus, Ohio, made half-track conversion kits for the Fordson in 1926. Besides increasing the ground-contact area, the tracks nearly doubled the weight of the Fordson, greatly improving traction.

110

1948 Ford Model 8N with Bombardier tracks

Above: *Bombardier, the Canadian aircraft and snowmobile maker, devised this half-track kit for the famous Ford tractor as well as other brands. The tracks were especially effective when a front end loader was employed. Half-tracked Fords were a favorite of maple-syrup producers because the tracks allowed the tractor to operate in deep snow. This 8N was also equipped with an auxiliary two-speed gearbox. Owner: Palmer Fossum.*

1958 Minneapolis-Moline Jetstar II crawler

Left and below: *Minneapolis-Moline built just fifty-one Jetstar III crawlers. Featured was a 206-ci (3,374-cc), four-cylinder engine; five-speed transmission; foot-controlled shuttle shift; and differential steering. Owner: the Keller family. It was restored by M-M expert Rex Dale.*

Holt, Best, and Caterpillar: The Industry Standard

Benjamin Holt's steam Caterpillar was the first successful agricultural crawler, although later versions found good use in logging operations. Originally, the Holt Manufacturing Company of Stockton, California, simply converted one of its wheeled steam traction engines to crawler tracks. Holt was quite experienced with link-belts, or flat chain drives, having pioneered their use in both its "traveling combined harvesters" and in the drive chains for its wheeled steam traction engines. Holt reportedly bolted blocks of wood to parallel link belts to make the first tracks for his 1905 experiment.

The first Holt steam Caterpillar was sold in 1906 to a Louisiana company working in the Mississippi Delta. Steam machines sold well, but by 1908, internal-combustion power was in vogue. Holt incorporated the Aurora Engine Company also of Stockton to build gas engines both for his crawlers and for other companies as well.

Development and acceptance of the gas Caterpillar came in large part from their use in building the great Los Angles Aqueduct, begun in 1908 and comparable in scope to the construction of the Panama Canal. Some twenty-eight Caterpillars were pressed into drayage duties along the 233-mile (373-km) aqueduct. Much of the route was across the Mojave Desert, which proved to be a severe test for the machines. Most importantly, the work forced the rapid development and improvement of the Holt machines.

The Daniel Best Agricultural Works and Benjamin Holt had been competitors in the combine and steam-traction-engine businesses since the 1880s, when in 1908 the seventy-year-old Best suddenly retired and sold out to Holt. As part of the deal, Daniel's son, Clarence Leo "C. L." Best, was to invest in the new joint company and hold a responsible management position. C. L. worked for the new firm for about two years, but he never paid for his shares, and his position never satisfied him.

In 1910, C. L. Best left Holt and started his own C. L. Best Gas Traction Company of Elmhurst, California—with financial help and blessings from his father. It started making gas-engine wheeled tractors and an almost direct copy of the Holt 75 Caterpillar crawler.

The two companies wrangled in and out of court for years. One of the main bones of contention was the basic crawler track patent. In a swift move, Best's attorney made a deal with Alvin Lombard, maker of the famous Lombard Log Hauler, to buy his patent rights which pre-dated Holt's crawler patents. Lombard was willing to sell as he believed Holt had pirated his design in the first place and then had refused to negotiate payment for the rights. Best prevailed in the lawsuit and forced Holt to pay royalties to continue to use the design.

Holt sold many Caterpillars to the Allied governments during World War I and was given credit—possibly more than he deserved—for helping develop the tank. Best did not get in on the government wartime largess, but came out of the war years with two classic crawler models—the Best Sixty and Thirty. These were uniquely designed and balanced machines that worked well and looked good. Best called his machines "Tracklayers" since the name Caterpillar had been trademarked.

Holt came out of the war with the 2-, 5-, and 10-Ton models; the later two bore the War Department's influence in their designs. While these were good

1919 Holt 10-Ton

The 10-Ton weighed just 9.5 tons (8,550 kg) and was rated by the factory at 75 hp. It was powered by a 929-ci (15,217-cc), overhead-valve, four-cylinder engine. This model was developed with the help of the U.S. Army in World War I. It had a unique segmented-track-roller frame, which gave it higher speed capability. It was also possible to "back out" of the tracks when reversing with a heavy load. Owner: the Vouk family.

1913 Holt Model 60

This Model 60 was purchased new for $4,205 by the Hahn brothers, who ran a ranch in Colusa, California. It was used on the ranch until it was sold at an estate sale in 1980. The 1,230-ci (20,147-cc), four-cylinder, OHV engine was rated at 50 drawbar and 60 belt hp at 500 rpm. Turns were made by releasing one steering clutch and pivoting the front tiller wheel. No steering brakes were provided, so only wide, sweeping turns could be accomplished. A considerable amount of human effort went into controlling one of these monsters. Owners: Larry Maasdam and Ron Miller.

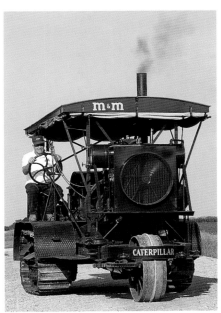

1921 Holt Model 75

The 75 was powered by a 1,400-ci (22,932-cc), four-cylinder engine and two-speed transmission. Bore and stroke was 7.50x8.00 inches (187.50x200 mm), giving 50 drawbar and 75 belt hp at 550 rpm. A K-W magneto provided spark. Besides the front tiller wheel, steering clutches and brakes were provided on the Model 75. The large wheels at the rear contained the clutches for the tracks. The machine weighed about 23,000 pounds (10,350 kg). Owner: Larry Maasdam.

and rugged tractors, the larger two of the three never seemed to garner the popularity of the Best machines. The Holt 2-Ton was, however, a favorite among both farmers and loggers.

In April 1925, in an effort to stem the endless patent and sales conflicts, Holt and C. L. Best combined forces as the Caterpillar Tractor Company. The financial houses brought this about, since there was not really enough business for both firms, and the endless struggle for patents was sapping too much energy. The Thirty, Sixty, and 2-Ton were continued, but Holt's 5-Ton and 10-Ton models were dropped after a short time. By 1927, the first all-new Caterpillar debuted as the Model Twenty, replacing the 2-Ton. Smaller and larger "Cats" followed, and ancillary equipment was added, such as the graders made by the Russell Grader Manufacturing Company of Stephen, Minnesota.

In 1931, Caterpillar unveiled the first diesel tractor—a seminal event in the history of tractors. After a grueling $1.5-million, two-year effort, the Diesel Sixty was tested successfully at the University of Nebraska. Despite many growing pains, the diesel set new standards for economy and rugged power. Although Caterpillar continued gasoline engines in ever larger sizes into the 1930s, the diesel engine ruled after World War II.

1925 C. L. Best Thirty Orchard
Above: *C. L. Best tractors used the trade name "Tracklayer" to distinguish them from the Holt Caterpillars. Upon introduction, the Thirty was originally called the Model S. It was built by C. L. Best and Caterpillar between 1921 and 1932, during which more than 23,000 were sold. Owner: Jerry Gast.*

1933 Caterpillar Ten
Left: *Caterpillar built almost 5,000 Tens from 1928 to 1932. The Ten was one of the few Caterpillars to use a side-valve engine, rated at 10 drawbar and 15 belt hp. Owner: Marv Fery.*

1935 Caterpillar Twenty-Eight
The Twenty-Eight was an update of the Twenty-Five, more closely reflecting its true power capabilities. On the drawbar, the Twenty-Eight was rated at 30 hp. Owner: Marv Fery.

1947 Caterpillar D4 Orchard
The D4 U Series featured a new 4.50x5.50-inch (112.50x137.50-mm) engine with cross-flow cylinder heads. The 350-ci (5,733-cc) diesel engine produced 421 foot-pounds of torque at 1,000 rpm. It normally used a two-cylinder, horizontally opposed starting motor with a 2.75x3.00-inch (68.75x75-mm) bore and stroke, but this example has direct electric start. Author Robert Pripps is at the controls. Owner: Larry Maasdam.

1930s Caterpillar Model Twenty-Two Orchard

Looking more like a twenty-first-century moon rover than a tractor from the 1930s, this Twenty-Two bears the complete set of factory orchard accouterments. The streamlined fenders were designed to let the tractor slip through low-hanging branches without getting caught. A four-cylinder, OHV engine with a bore and stroke of 4.00x5.00 inches (100x125 mm) powered the crawler. Owner: Larry Maasdam.

Best/Caterpillar Sixty

C. L. Best Gas Traction Company of Elmhurst, California;
Caterpillar Tractor Company of Peoria, Illinois

C. L. Best's Sixty was introduced in 1919 and tested in Nebraska in 1921. Despite being a popular tractor noted for pulling power and durability, Best had considerable trouble getting it through the tests. This was mainly the fault of the engine not living up to its expectations. The four-cylinder, 1,128-ci (18,477-cc) engine just could not be coaxed to make 60 hp. With the air cleaner removed and the engine operated at 656 (rather than 650) rpm, 56 hp was the most it could do.

In 1923, Best was back at Nebraska with the engine improved by cam and valve changes. Now, at 650 rpm and with the air cleaner in place, the Sixty was able to produce 66 hp. Finally, the official test backed what loyal owners had known all along.

The Sixty was equipped with a two-speed transmission, but a three-speed unit was optional. Later, the three-speed box was standard. The Sixty originally weighed 17,500 pounds (7,875 kg), but was increased to more than 20,000 pounds (9,000 kg). Early versions had a low seat set far aft, with the steering levers out to the right side. This was fine for agriculture and construction, but for logging, a seat, which we would now call "conventional," was substituted. This logging, or cruiser, seat finally became standard. Selling price began at about $4,000 and rose to $6,000 by the end of production in 1931. Almost 20,000 were sold.

In 1931, a new, diesel engine was installed in the basic Sixty running gear and 157 Diesel Sixtys were made that year. With it, Caterpillar had the first production diesel tractor and pioneered the use of the pony-motor starter. In 1932, it was modernized as the Diesel Sixty-Five.

1920s Caterpillar Sixty brochure
The cylinder above the fender was the massive fuel tank, which a hard-working Sixty could easily empty in a ten-hour day.

1926 Caterpillar Sixty

In its heyday, the Sixty was considered the industry standard for large crawlers by farmers and loggers. Introduced in 1919 by C. L. Best, the Sixty became the Caterpillar Sixty after the 1925 merger. It overwhelmed the competition because of its reliable, well-balanced design. Owner: Dave Smith.

Cletrac-Oliver: "Geared to the Ground"

The White family of Cleveland, Ohio, was an inventive lot. Thomas White, the father, founded a sewing machine company in 1859 that bears his name to this day. The company ventured into transportation around the 1900s by manufacturing both roller skates and bicycles. Sons Rollin, Windsor, and Walter built the successful White Steamer automobile in 1900 and later converted their car line to use their own gasoline engines. In 1911, they started the Cleveland Motor Plow Company, later called the Cleveland Tractor Company. Their crawler tractors were named Cletracs and bore the slogan "Geared to the Ground."

Cletracs were unique in that they employed what was then known as "differential steering." Simply, on a Cletrac, engine power was delivered to a differential, as is the case with a wheel vehicle. To steer a Cletrac, one of the tracks was braked by means of a steering brake lever. Other brands of crawlers did not have a differential between the tracks, but employed a clutch and brake for each track, in addition to the normal clutch used for disengaging the whole powertrain from the engine. To turn, one track was declutched, and then that track was braked as necessary while the other track powered through the turn. Cletrac claimed the differential feature as a great benefit, although many operators found Cletracs harder to control than other brands.

Originally, Best Tracklayers were built with differential steering. At one point, when C. L. Best was having a hard time of it due to his inability to get materials to build tractors during World War I, Rollin White was able to gain control of the C. L. Best Traction Company by buying the stock of disgruntled stockholders. It was not long before Best was able to regain control, but in the meantime, his patent rights to the differential-steering concept were lost. After that, Best machines used the clutch-steering method.

These days, the term "differential steering" has a whole different meaning. Today, the most advanced crawlers employ a hydrostatic drive (hydraulic pump-motor) to bias the gear differential between the tracks,

1917 Cleveland 20

Founded by the White brothers, the Cleveland Motor Plow Company became the Cleveland Tractor Company in 1917 and then Cletrac in 1918. This diminutive crawler had a 12/20 rating provided by a Weidley four-cylinder engine. Owner: Charles Doble.

1917 Cleveland advertisement
"Geared to the Ground" was Cletrac's famous motto.

1920 Cletrac Model F
This 1,900-pound (855-kg) crawler demonstrated a drawbar pull of 90 percent of its own weight. It rode on an inner track of rollers that gave it a nice, even footprint. It was also one of the first to use the elevated drive sprocket.

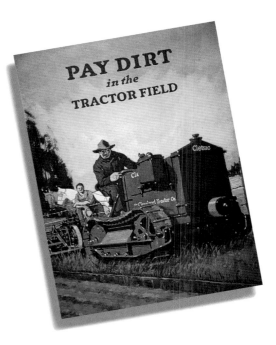

1920s Cletrac brochure

adding speed to one side and subtracting it from the other. By simply moving a hydraulic metering pin, one track can be made to go forward while the other runs in reverse for a pivot turn. Thus, a Caterpillar Challenger, for example, can be spun around in its own length by just an easy movement of the steering wheel.

The pros and cons of differential steering not withstanding, the Cleveland Tractor Company made a credible line of crawlers from 1916 through 1944, when it was sold to the Oliver Corporation. Then, the designs generally continued with little change through the end of the Oliver name. Remarkably, it was the White Motor Corporation that bought Oliver in 1960. In 1965, the crawler line was discontinued.

The smallest Cletrac was the 9-hp, 1,890-pound (850-kg) Model F built from 1920 to 1922. It was characterized by high drive sprockets that resemble modern Caterpillars. During the 1930s, Cletrac went head to head and model for model with Caterpillar. The Cletrac 80 even bested the Cat D8 with 96.73 hp versus the D8's 95.27 hp. The largest Oliver Cletrac was the 130-hp OC-18 of 1952. The OC-18 used a six-cylinder, 895-ci (14,660-cc) Hercules diesel engine.

1936 Cletrac CG

The CG was not a small tractor, weighing almost 6 tons (5,400 kg). It featured a six-cylinder Hercules engine that developed a maximum of 50 hp, and a three-speed transmission. It also offered electric starting.

1952 Oliver HG

The HG had been introduced by Cletrac in 1939. It was powered by an L-head, four-cylinder Hercules engine. It weighed about 2 tons (1,800 kg). Oliver bought Cletrac in 1944 and continued production of the crawler line with Oliver livery. In 1960, a re-organized White—erstwhile founder of Cletrac—bought the Oliver Corporation, proving that what goes around comes around.

Monarch and Allis-Chalmers: Persian Orange Crawlers

The company named after Edwin P. Allis and William J. Chalmers was founded in 1901. Allis was the largest builder of industrial steam engines in the United States. Chalmers was president of Fraser & Chalmers, manufacturers of mining equipment. Allis had died a few years before the merger of the two firms into the Allis-Chalmers Company of Milwaukee, Wisconsin, but he was well represented by his two sons, Will and Charles, and by his nephew Edwin Reynolds, who held the title of chief engineer in the new company.

Almost from the beginning, things did not go well for Allis-Chalmers. The economy was not conducive to the sales of their products, and the board of directors could not pull together to take corrective action. By 1912, the firm went into receivership, with Delmar Call and Otto Falk appointed receivers. The company was then reorganized as the Allis-Chalmers Manufacturing Company. Falk, a general in the Wisconsin National Guard, was named president, a position he would hold until 1932.

Falk saw that the diversification efforts of his predecessors had failed because all eggs were in the same heavy-capital-equipment basket. When the economy was down for one branch, it was down for all. Falk determined that

1939 Allis-Chalmers Model S
This Model S was equipped with a nice cab from its former duties plowing snow. The 10-ton (9,000-kg) S used a 675-ci (11,057-cc), four-cylinder engine developing a maximum of 84 hp. Featured was five-speed transmission giving a top speed of over 6 mph (9.6 kmh).

A-C should go into the agricultural-tractor business with lightweight, inexpensive tractors. Undaunted that his first efforts were not successful, Falk pressed on into larger, more Fordson-like machines by 1918.

The new Allis 15/30 and 18/30 did better, but the tractor business was still loosing money. After a ten-year attempt, Harry Merritt was hired in 1926 as tractor manager, ostensibly to close out the tractor business. Instead, Merritt, a real driver, revitalized the effort, cutting the cost of the tractor line almost in half by re-engineering every part. Then, in 1928, Merritt bought the Monarch Tractor Corporation of Springfield, Illinois. In 1929, he came up with the bright Persian orange paint color. It was the first bright-colored tractor—and essentially became the line's trademark. The addition of Advance-Rumely in 1931, with its dealer and branchhouse network, put Allis-Chalmers in the business for the long haul.

The Monarch line, at the time of the acquisition by A-C, included two crawlers, the Models F and H. Merritt changed the designations to the Models 75 and 50, respectively, to reflect their drawbar horsepower capability. The Monarch label was retained for several more years.

The first original Allis-Chalmers crawler was the Model 35 of 1929. It too was originally called a Monarch, but by 1933, number designations were changed to letters, and the 35 became the K. It was at about that same time that the Monarch moniker was dropped, as was the steering wheel in favor of conventional levers. A-C crawlers had steering clutches and brakes and a hand master clutch.

In the mid-1930s, A-C took a diversionary route into the realm of the semi-diesel for its crawler line. Fuel oil was injected into the engine, but the compression ratio was not high enough to set off the charge. Therefore, a magneto and spark plugs were used. The semi-diesel did not have the fuel consumption advantages of the pure diesel and was never well accepted. It was dropped in 1940 when the HD-14 debuted with a 6-71 two-cycle GM diesel.

A-C used GM diesels in its line of crawlers until 1955 then came out with four-cycle diesels. The HD-21 was the first, with a turbocharged, 844-ci (13,825-cc) six. The largest A-C crawler was the 1970 HD-41, a 50-ton (45,000-kg) monster that had to be assembled on the job. It had a 524-hp engine and could be equipped with a 30-foot-wide (9-m) dozer blade.

In 1974, a merger of Allis-Chalmers and Fiat of Italy created the Fiat-Allis Corporation.

1934 Allis-Chalmers Model K
Originally called the 35, this tractor was later labeled the Model K. Production of the Model K, albeit without the steering wheel, was continued through 1943. A Model K-O version used Allis-Chalmers's semi-diesel engine, a low-compression diesel that used spark plugs to set off the charge. The four-cylinder engine displaced 461 ci (7,551 cc). Owner: Alan Draper.

1969 Allis-Chalmers HD-4

Above and right: *The HD-4 was marketed from 1966 to 1969, with a selling price of $12,000 including the bulldozer blade. The A-C Model 2200 naturally aspirated four-cylinder diesel engine displaced 200 ci (3,276 cc).*

1939 Allis-Chalmers Model L

Facing page, top: *The 22,000-pound (9,900-kg) Model L was powered by a six-cylinder engine of 844 ci (13,825 cc) that delivered more than 90 hp. The twin-carburetor engine featured a split exhaust manifold with twin exhaust pipes. The L was built from 1931 through 1942. Owner: Norm Meinert.*

1941 Allis-Chalmers Model M

Facing page, bottom: *The M was introduced in 1932 and built through 1942. The engine was rated at 32 hp and was the same unit as used in the Allis-Chalmers Model U wheeled tractor.*

1945 John Deere Model BO-Lindeman
John Deere Model BO Orchard tractors were converted to crawler tracks by the Lindeman brothers of Yakima, Washington. The tractor featured the famous John Deere two-cylinder engine of about 25 hp. As converted, the crawler weighed 5,500 pounds (2,475 kg). Individual steering clutches were added. Owner: Harold Schultz.

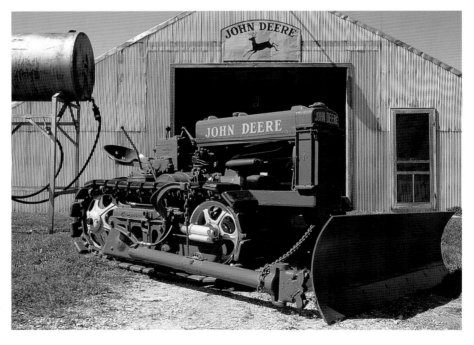

John Deere Model BO-Lindeman

Lindeman Power Equipment of Yakima, Washington;
Deere & Company of Moline, Illinois

Deere got into the crawler market almost by accident. It all started in 1923 in Yakima, Washington—far afield from Deere's headquarters in Illinois—where Jesse and Harry Lindeman bought the implement business where Jesse had been working. They became a Holt Caterpillar dealer and sold crawlers to their orchard customers. In 1925, when Holt and Best merged, the local Best dealership had seniority, and the Lindemans were out. They then sold the Cletrac and John Deere lines.

Neither the Cletrac crawlers nor Deere's Model D or new Model GP wheeled tractors were really suited for orchard work as they were too high to fit under the trees and had vertical protrusions, such as exhaust pipes and air inlet pipes, that would be knocked off by low-hanging limbs. The brothers modified a GP for orchard work by cleaning up the protrusions and making special castings to lower it. The folks at Deere were impressed. They adopted the Lindeman's ideas, and the first Deere orchard tractor was born—the Model GPO.

In 1932, the brothers again told Deere they had something new to show them. They had adapted the tracks from a Best Thirty to a Deere Model D. Thorough testing proved the concept, but a decision was made to discontinue the D crawler and build a crawler version of the GPO, which was receiving acclaim from orchard owners.

The GPO-Lindeman was built from 1933 until 1935. At first, the usual individual brakes were used for differential steering. Later, the differential was removed and steering clutches incorporated for improved control. In 1935, production of the Model GP was halted in favor of the new Model B. An orchard version, the Model BO, was included in the line. The Lindeman brothers went right on receiving partially completed tractors from Deere and finishing them with tracks, creating the Model BO-Lindeman of 1939–1947.

Deere was so impressed by the Lindeman brothers' work that it bought their operation on January 1, 1947. Later in the year, the Model M replaced the Model B, and Lindeman continued with the conversions in Yakima, creating the MC (M Crawler). In 1953, the updated 40C replaced the MC.

In 1954, Deere moved the crawler operations and many of its employees from Yakima to Deere's Dubuque, Iowa, factory that had been producing the basic tractors that supplied the Yakima plant, a means of consolidating operations for future expansion. Deere planned to get into construction/industrial machinery in a big way. Since then, Deere has produced a full line of thoroughly modern crawlers, including rubber-track versions specifically for agriculture.

1940s Lindeman Model BO-L leaflet

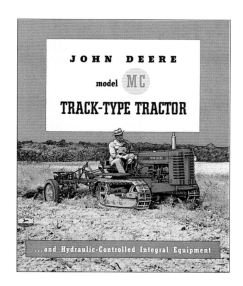

1950 Deere Model MC brochure

The MC replaced the BO crawler. Tractor base units were shipped from Deere to Lindeman in Yakima to have tracking fitted for the West Coast market, and track units were shipped to Iowa for the rest of the United States and export. The new model soon found favor not only with fruit farmers in the hilly Northwest but also with housing and other small contractors. The last MC was built in 1952.

Orphan Crawlers

1915 Bullock Creeping Grip and Neverslip advertisement

The Bullock Tractor Company of Chicago termed its machine the "Flat Wheel Tractor" and promised that every ounce of power was utilized. Even with its evocatively named machines, Bullock suffered several bankruptcies before finally exiting the field in the 1920s.

As with the tractor industry as a whole, it was survival of the fittest in the specialized world of the crawler tractor. Numerous small firms attempted to build their own crawlers in the pioneering years of the gasoline tractor, and many of these machines ended their days as orphan crawlers, wiped out by the economic recession following World War I or the Great Depression of the 1930s.

1917 Bates Steel Mule Model D and C advertisement

"A Tractor Now for All Conditions," announced this ad from the Joliet Oil Tractor Company. The colorfully named Bates Steel Mule went through numerous permutations over the years, from the lightweight 15/22 Model D to the massive Model 80. Joliet merged in 1919 with the Bates Tractor Company of Lansing, Michigan, creating the Bates Machine & Tractor Company. The Great Depression forced production to end in 1937.

Yuba Ball Tread 12/20, 1916

Rather than bogie wheels, the miniature Yuba 12/20's track system rode on large ball bearings between the track and track frame. The Yuba Construction Company of Marysville, California, bought rights to the tractor in 1914 from the Ball Tread Company of Detroit. Ball Treads were made in sizes up to 40/70, always with a tiller wheel. The name was changed to Yuba Manufacturing Company in 1918, but the firm disappeared during the Great Depression.

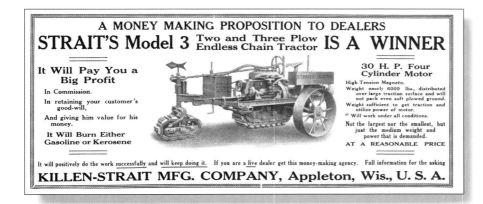

1916 Strait's Model 3 advertisement

The Killen-Strait Manufacturing Company of Appleton, Wisconsin, offered its "Endless Chain Tractor" in the 1910s. Oddly enough, the Strait's Model 3 featured a single wheel to offset the tracks.

1919 Pan Tank-Tread advertisement

The Pan Motor Company of St. Cloud, Minnesota, began to build automobiles in 1916 before introducing its crawler tractor in 1917. The Pan firm's founders were better at raising money than building and selling cars or crawlers, however, and questionable business practices and legal wrangling quickly sank the company.

Caterpillar D8

Caterpillar Tractor Company of Peoria, Illinois

1936 Caterpillar RD8
The RD8 was "big iron" in its day as it featured a six-cylinder engine of 103 hp. The Diesel Seventy-Five became the RD8 in 1936, and the "RD" designation was carried through 1938; the designation was simply "D8" thereafter. The RD8 was a big machine, weighing almost 34,000 pounds (15,300 kg). Owner: Larry Maasdam.

When push comes to shove, Caterpillar's D8 was almost everybody's favorite crawler. It was the Boeing B-52 Stratofortress of tractors until the D9 came out in 1954. By then, the term "D8" was, like the B-52, a synonym for overkill. Nobody really needed more power than the D8 offered—although buyers of the D9, D10, and D11 seemed to think they did.

The D8 was an outgrowth of the Diesel 70 of 1933. In late 1933, Caterpillar created a standard engine with a bore-and-stroke configuration of 5.25x8.00 inches (131.25x200 mm) but using three, four, and six cylinders. When the standard six-cylinder engine was put in the Diesel Seventy chassis, it became the new Diesel Seventy-Five. In 1936, Caterpillar changed its model designators that tended to signify horsepower to a new letters-and-number system that had no such connotation. The Diesel Seventy-Five became the RD8. Later, "R" designators were used for gasoline tractors and a "D" indicated diesel, hence, the D8.

The most famous of the classic D8s was the U Series of 1947–1955. The standard engine was now 5.75x8.00 inches (143.75x200 mm), and power in 1955 was up to 155 hp. The D8 weighed 37,000 pounds (16,650 kg). A five-speed transmission was used.

Stories abound of the legendary feats of the D8. Some come from the author's uncle, Norman Pripps, a World War II Seabee in the South Pacific. His outfit had Allis-Chalmers and International crawlers in addition to D8s. They left the others behind at the end of each island campaign, but kept the same D8s throughout the war. The others required a push to fill a LeTourneau scraper; the D8 could fill the same scraper to overflowing without assistance. When pushing over trees with the D8, the bulldozer blade could cave in if the tree was stubborn.

The D8 had the right balance, power, and a pleasing exhaust note. There is a story of a driver on a logging job that was given an International to operate in place of his aging D8. When asked why he had ripped off the muffler, he responded, "If I can't drive a Caterpillar, at least I can drive something that sounds like one."

1934 Caterpillar Diesel Seventy-Five
The Diesel Seventy-Five was made from 1933 to 1935 and boasted 93 hp from its 5.25x8.00-inch (131.25x200-mm), six-cylinder engine. Owners: the Skirvin Brothers; Carl Skirvin is at the controls.

CRAWLERS GO TO WAR

"No one can see far ahead in these days, but of this we can be certain: Mechanical devices for winning battles will be the predominant factor. Brave men will still be essential to the proper handling of war machines, but it will be a war of machinery, rather than a war of flesh."
—French World War I army commander Marshall Ferdinand Foch, 1926

Farmer Leo Steiner, owner of a large agricultural estate in Hungary, read reports of the large-scale mechanized farms in the United States. Because much of his estate had wet soil, Steiner decided to order a Holt Caterpillar Model 60. He liked the machine so much that he applied to be a Holt dealer, and in 1912, he was awarded the dealership for the countries of Hungary, Austria, and Germany.

Steiner then challenged all competitors throughout his territory to pulling and plowing contests. The prowess of the Caterpillars in these contests soon caught the attention of the Austrian military. The Austrians asked for further demonstrations and invited their German military counterparts to witness them as well. The German representatives were not convinced of the Holt's military significance, but their opinion changed in 1913, and Steiner was ordered to import all the Caterpillars he could get, and to look into manufacturing rights. Fortunately for the Allies, the Germans had delayed their decision long enough for World War I to break out, and all trade with the Central Powers came to a halt before more than a few Caterpillars were delivered.

On June 28, 1914, the Crown Prince of Austria was assassinated in the city of Sarajevo, Serbia, and war broke out between Austria and Serbia. Due to their mutual defense treaties, Germany, France, and England were dragged into the conflict on both sides, sparking World War I. The United States was drawn in to the conflict on April 6, 1917.

Orders for Holt Caterpillar 75s quickly followed from several of the Allies. After Britain, France, and Russia had ordered more than 1,000 Caterpillars, the U.S. War Department finally consented to tests in 1915. The War Department still held to the old philosophy that only animal power was reliable enough for war use, but in 1916, the U.S. Army finally purchased twenty-seven Holt Caterpillars.

The Army Caterpillars were soon pressed into service when President Woodrow Wilson ordered General John "Blackjack" Pershing into Mexico in a punitive foray against Mexican revolutionary Pancho Villa. While Pershing's expedition failed to capture Villa, Caterpillar tractors were instrumental in conveying troops and supplies some 350 miles (560 km) into then-roadless Mexico. Resounding praise by Pershing for the Caterpillars gave them credibility with the army.

Meanwhile, British Army Colonel Earnest D. Swinton heard reports of the performance of the Holt Caterpillars at the front in Belgium: The big crawlers were virtually unstoppable in even adverse conditions. He had the British Foster Company build a completely armored tracked vehicle—soon known as a "tank"—using the Hornsby-Akroid-Roberts track technology previously sold to Holt. These tanks were first used in September 1916 in the Battle of the Somme in France with great effectiveness, putting an end to the concept of trench warfare.

While Holt had nothing to do with the tank, Caterpillar techniques were employed. At the same time, Caterpillars were being armor-plated by the Allied countries to enable them to pull big guns into firing positions. As a result, American newspapers gave credit to Holt for the development of the tank. After the war, Swinton lavished praise on Holt, calling the Holt works in Stockton "the cradle of the tank."

1940s Caterpillar advertisement
From its role as the forefather of the tank in World War I, the crawler served patriotically in World War II in all theaters. Caterpillars did everything from helping to build the Alcan Highway to clearing beachheads with the Navy Seabees.

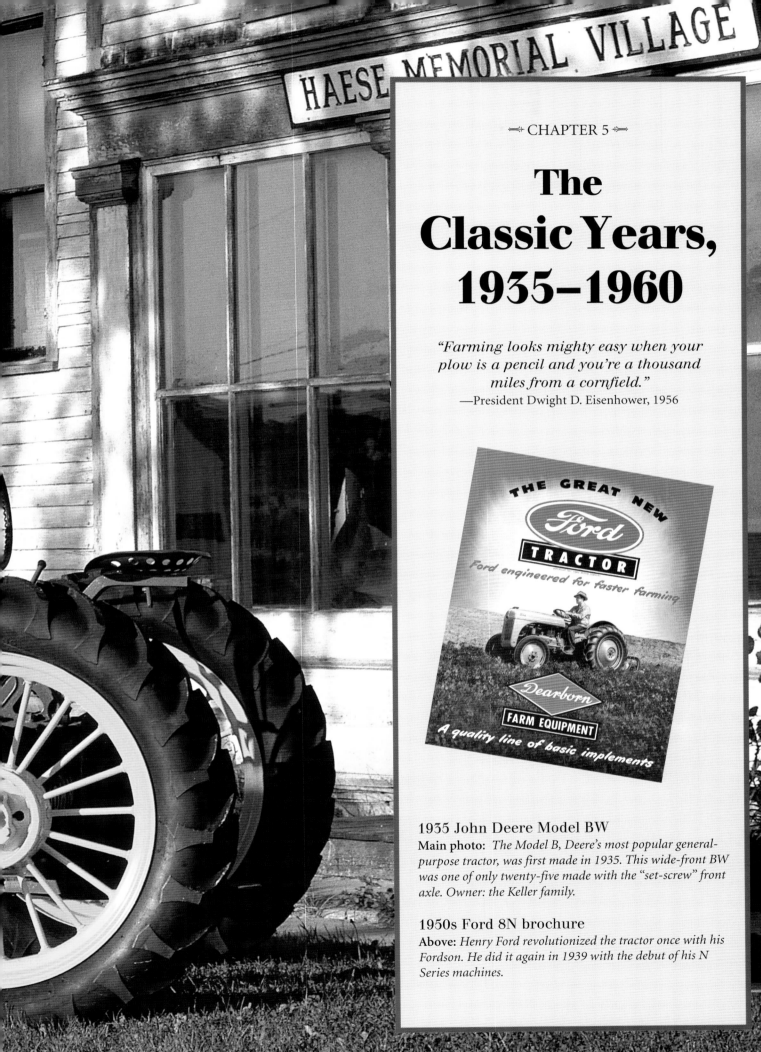

CHAPTER 5

The Classic Years, 1935–1960

"Farming looks mighty easy when your plow is a pencil and you're a thousand miles from a cornfield."
—President Dwight D. Eisenhower, 1956

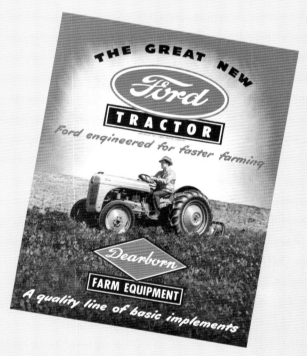

THE GREAT NEW
Ford TRACTOR
Ford engineered for faster farming
Dearborn FARM EQUIPMENT
A quality line of basic implements

1935 John Deere Model BW

Main photo: *The Model B, Deere's most popular general-purpose tractor, was first made in 1935. This wide-front BW was one of only twenty-five made with the "set-screw" front axle. Owner: the Keller family.*

1950s Ford 8N brochure

Above: *Henry Ford revolutionized the tractor once with his Fordson. He did it again in 1939 with the debut of his N Series machines.*

The Best of Times and the Worst of Times

Timeline

1937: Dirigible Hindenburg crashes in New Jersey

1939: Germany invades Poland; WWII begins

1939: First American TV broadcast

1941: Japan bombs Pearl Harbor; United States enters World War II

1944: Allies invade France on D-Day

1945: V-E Day May 7; V-J Day August 15, after two nuclear bombs dropped on Japan

1946: U.S. Air Force Captain Chuck Yeager is first man to exceed the speed of sound

1950: Korean War begins

1951: First hydrogen bomb exploded

1954: *Nautilus*, the first nuclear submarine, launched

1959: First computer chip patented

By 1935, the United States had passed through the worst of the Great Depression. Tractor sales were rising, and the devastating drought of the first few years of the 1930s was over. While things were still tough both on the farm and in the city, President Franklin Delano Roosevelt's New Deal policies were at last having some of the desired effects. Roosevelt's programs had helped increase farm income. In some rural areas, electricity was available; some even had telephone service. Things were looking up for the farmer.

The growling Fordson had been banished to England by the new, general-purpose, row-crop machines. Steam was all but gone from the scene. Tractor sales for 1934 were up 40 percent over 1933, and the surviving tractor makers were offering new, modern models.

Dark specters were lurking on the horizon, however. Minatory dictatorships had risen from the ashes of World War I, which had ended a mere eighteen years earlier. Adolph Hitler's National Socialism threatened to engulf Europe, drawing the globe into its second world war. The farm crawler's military sibling, the tank, became one of the key strategic weapons on both sides of the battle. Afterwards, the remaining tensions of World War II sparked the Cold War.

The 1940s and 1950s were the best of times and the worst of times for the farm tractor. The market for tractors grew steadily and dramatically until 1954, when another industry shakeout similar to that of the 1920s resulted in a market consolidated in the hands of just a few manufacturers.

Throughout these decades the basic shape of the tractor changed little from the general-purpose machine personified by International Harvester's Farmall. Developments were more evolutionary than revolutionary, however, including the rise of live hydraulics, new fuels, and independent PTOs. Tractor makers continued to build a better tractor, thanks in large part to engineering lessons learned during World War II.

1937 Massey-Harris Challenger
Above and facing page: *The Challenger was a modernized row-crop version of the Wallis 10/20 and was built from 1936 to 1938. Owner: Howard Dobbins.*

VICTIMS OF THE SCRAP HEAP

1942 Ford-Ferguson 2N
What every tractor collector hopes to find: a forgotten classic hiding in the woods and available for sale.

During the early stages of World War II, many changes took place to get the entire civilian population on a wartime footing. Men too old or physically unfit for the service, as well as women of all ages, were urged to get employment in "defense" plants, on farms, or in construction for the war effort. Gasoline, butter, meat, shoes, and many other items were rationed. Each family member got a ration book with tear-out coupons, and a gas ration book was issued for each vehicle.

At the same time, the Boy Scouts, Girl Scouts, churches, VFW posts, and junk dealers held scrap-metal drives. Anyone who had an old or disabled tractor or other vehicle was encouraged to donate it to the war effort. Both of the author's grandfathers donated to the war scrap metal drive—one a fine old Case steam engine, the other a Cleveland-made 1918 JT crawler. An untold number of the oldest tractors met their fate in that way.

1940 Oliver 80 Diesel
Left and below: *The 80 was an outgrowth of the Oliver Hart-Parr 18/27 and 18/28 tractors of 1930–1937. The diesel version was introduced in 1940 with a Buda-Lanova engine. Later, Oliver replaced the engine with one of its own design.*

1935 Allis-Chalmers Model UC
Right: *The UC was A-C's first row-crop tractor and was otherwise the same as the Model U. It was built from 1930 to 1941. Originally equipped with a Continental engine like the Model U, later versions had an Allis engine.*

1941 Case Model VC
Case's Model VC replaced the Model RC in 1940 and was replaced by the VAC in 1942. The four-cylinder L-head engine of the VC was supplied by Continental.

1955 Ferguson TO-35
After the split with Ford, Harry Ferguson's TO-35 was built in Detroit. It had a 134-ci (2,195-cc) Continental engine, six-speed transmission, and weighed about 3,000 pounds (1,350 kg). Owner: Palmer Fossum.

Scrap metal drive
During World Wars I and II, scrap metal drives melted down many a classic farm tractor, turning plowshares into swords for the patriotic war effort. This 1940s ad from Minneapolis-Moline warned farmers that liberty—in the form of the Statue of Liberty—was in danger if they didn't harvest their old tractors for the war. Unstated, naturally, was that farmers would then need a new tractor, preferably a Prairie Gold Minne-Mo.

Deere Model B
Deere & Company of Moline, Illinois

John Deere's Model B probably introduced more farmers to power farming than any other tractor since the venerable Fordson. From its introduction in 1935 until its retirement in 1952, well over 300,000 Model B tractors were delivered.

The B was designed more or less simultaneously with the larger John Deere Model A, which made its debut in 1934. Both had a hydraulic implement lift, adjustable rear-wheel tread width, and one-piece rear-axle housings—new features in 1934–1935. The 2,800-pound (1,260-kg) Model B was designed to compete with a team of horses and cost not much more. In fact, since horses tired and had to be rested, the original B could do the work of two teams. Later-year models became more powerful and heavier, eventually supplanting the original A.

The original Model B two-cylinder engine displaced 149 ci (2,441 cc) and produced about 15 hp on kerosene via a four-speed transmission.

1930s John Deere Models A and B brochure

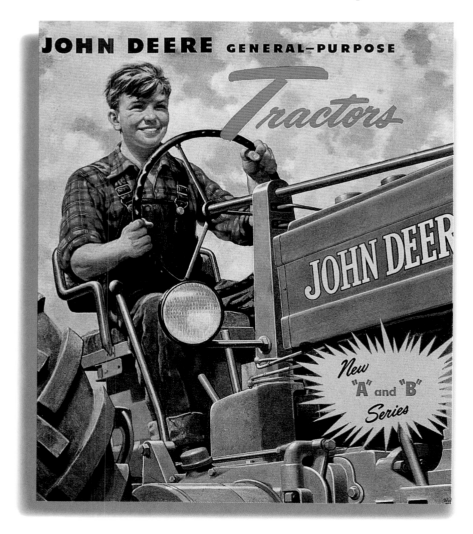

1935 John Deere Model B-Garden

This B, serial number 1798, was built on December 12, 1934. After April 1935, production front pedestals were changed from four to eight bolts, and the designation changed to BN (narrow) indicating the single front wheel. As acquired by Walter Keller in Helena, California, this tractor had the wrong front end. Keller found the right parts and disassembled the front, propping it up on blocks. While working on it, he decided to change the leaking flywheel seal at the same time. In the process, he rolled the flywheel a quarter turn, and the engine fired up. Needless to say, there was a scramble to shut it off.

In 1938, the B received Henry Dreyfuss's styling treatment and a 19-hp engine of 175 ci (2,867 cc). In 1947, the B was restyled yet again by the Dreyfuss studio. Engine displacement for these "late-styled" B's was increased to 190 ci (3,112 cc), producing some 28 hp in the gasoline version with a six-speed gearbox.

The Models B and A were the first of Deere's new general-purpose tractors. Dual narrow front wheels were the conventional configuration, but single-front-wheel and wide-front versions were available, along with standard-tread, industrial, and orchard models. Later, high-crop and special narrow adaptations of the basic B were offered, as was the Lindeman crawler version of the orchard B.

1935 John Deere Model BW

Below: *On the right is the B's owner, Walter Keller. On the left is eighty-three-year-old Ruben Schaeffer, owner of the barn behind the tractor.*

1935 John Deere Model Bs

Above: *Two rare Deere Model B tractors with two famous collector-restorers at the controls. On the left is the BW with Walter Keller. On the right is a B-Garden with Walter's son Bruce.*

1935 John Deere Model BW
This wide-front BW has the optional F&H wheels and rubber tires. Early Model B tractors, such as this one, used a 149-ci (2,441-cc), two-cylinder engine.

1943 John Deere Model BN Styled

This wartime Model BN (narrow, single front wheel) was delivered on steel. Road gear was not included by Deere when tractors were not shipped on rubber tires. From 1941 to mid-1947, the B was equipped with a 175-ci (2,867-cc) engine; after that, a 190-ci (3,112-cc) engine was used. Owner: Larry Maasdam.

Streamlined Styling Comes to the Farmyard

In the middle of the twentieth century, manufacturers suddenly came to the startling realization that visible product differentiation had a positive effect on sales. Dramatic colors and stylish lines flourished on everything from household appliances to food packaging to home radios. Cars, which had been pretty much alike, now began to sport individualistic radiator grilles, sweeping and skirted fenders, raked windshields, and engines with as many as sixteen cylinders.

The radiator grille and the louvered hood side panels were the hallmarks of automotive styling in the early 1930s, so it was little wonder that these items found their way to the tractor. The first volume-produced American tractor to get this treatment was the famous Oliver Hart-Parr 70. It was billed as being so "car-like" that "Bud" and "Sister" could take their turns at the wheel.

The word "streamlined" was soon being used to describe new tractors from a variety of manufacturers. The 1936 orchard version of the Model J from the Minneapolis-Moline Company of Minneapolis looked like it could take its place at the starting grid of the Indianapolis 500 race. In 1937, stylish new sheet metal graced Huber's Model B, Massey-Harris's Pacemaker, Minneapolis-Moline's Z, and the glorious Graham-Bradley from the Graham-Paige Motors Corporation of Detroit, Michigan.

In 1938, the big players in the league reacted. Deere hired noted New York industrial designer Henry Dreyfuss to style its entire line of tractors. The Dreyfuss firm has left its mark on the design of Deere tractors ever since. International Harvester hired the famous automobile stylist Raymond Loewy. The first of the beautiful red Internationals was the long-lived TD-18. These were more than just sheet metal coverings over old machinery: These two design experts analyzed everything from a proper seat to adequate visibility for cultivating, making real and lasting improvements.

The zenith of avant-garde styling came in 1938 on the Minneapolis-Moline Comfortractor Model UDLX. Equipped with the first "factory cab," the UDLX had all the amenities of a sports coupe of the day. It was promoted as a tractor that could be

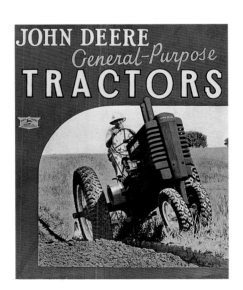

1940s John Deere Models A and B Styled brochure

worked all day and driven to town in the evening at its top speed of 40 mph (64 kmh).

The influence of automotive styling and engineering on tractors reached a peak in 1938 when Henry Ford began work on what was to become the Ford-Ferguson 9N tractor. Ford assigned his Ford Motor Company styling department to come up with the layout. The look of the new tractor complemented the rest of the Ford, Mercury, and Lincoln car and Ford truck lines, which were then classics of art deco. Most of all, the design was functional, could be produced economically, and was sized so that tractors would fit crosswise in a boxcar for maximum shipping economy.

Tractor styling lasted ten to twelve years before refreshing was required. Color became an identifiable trademark for tractor makers, and few changed colors after the 1940s. Toward the end of the 1950s, ergonomics—the science of human convenience—began to be considered. Such things as handholds and safety shields were added, along with better mufflers, lights, and convenient controls. Finally, towards the end of the 1950s, factories began experimenting with insulated, soundproofed, and air-conditioned cabs.

1947 John Deere Model G Styled

The G was introduced in 1937 as the row-crop equivalent of the standard-tread Model D. Power was about the same, although the engines were different. The G was also about 1,000 pounds (450 kg) lighter than the D. The 1942 version received streamlined sheet metal crafted by famed industrial designer Henry Dreyfuss. The 1947 model saw the G in its final form with the backrest seat and electric starting.

1940s Oliver Model 66

Oliver's stylish 66 debuted in 1947 as a 1948 model. Gasoline, diesel, and LPG engines of 129 ci (2,113 cc) were offered, as well as a 144-ci (2,359-cc), distillate-burning engine. All were four-cylinders and produced around 20 hp. Standard-tread and row-crops, both wide and narrow front, were offered. This 66 was a wide-front row-crop diesel.

1941 Farmall M
Above, both photos: *The big M was the top of the Farmall line from 1939 to 1954, replacing the F-30. The M used a 247.7-ci (4,057-cc), four-cylinder engine and could handle a three-bottom plow. It was available in distillate, gasoline, or diesel versions. Sales of the M averaged more than 22,000 units annually.*

1937 Minneapolis-Moline YT
Minneapolis-Moline wanted a small tractor to compete with the Farmall F-14 and the Deere H, and believed the YT to be the answer. The YTs made were prototypes that never went into production, and this rare M-M tractor was one of only twenty-five made. It was unique in that it has a vertical two-cylinder engine of 93 ci (1,523 cc); the engine was basically the back half of a Z engine. Owner: Walter Keller.

Minneapolis-Moline Model UDLX Comfortractor
Minneapolis-Moline Company of Minneapolis

The most famous of the U Series was the UDLX, also known as the U-Deluxe and Comfortractor. This was designed to be a tractor that farmers could drive to town after it had spent the day working in the field, and top speed was a blazing 40 mph (64 km/h). The UDLX featured items like a shift-on-the-fly five-speed transmission, windshield wipers, high- and low-beam headlights, taillights, cigarette lighter, heater, speedometer, and seating for three.

Under the skin, the tractor was basically a Model UTS. While the enclosed cab was comfortable, the lack of hydraulics meant the back door had to be

1930s Minneapolis-Moline
UDLX brochure

kept open in order to reach implement levers. The tractor was not practical for other than pulling jobs, as there was no belt pulley or PTO. The UDLX was less than optimum on the highway as well, since it did not have sprung axles. Even at low speeds, it tended to waddle like a duck. Where the UDLX really shone was in the service of the custom combiner. The long-distant pulls between jobs could be made in relative comfort and at reasonably high speeds.

The UDLX used the UTS's four-cylinder, overhead-valve engine of 284 ci (4,652 cc) and about 40 hp. It weighed about 4,500 pounds (2,025 kg).

On the downside, without hydraulics the implements had to be hand lifted through the back door, there was no provision for a belt pulley or a PTO, and the lack of springs made the ride barely tolerable at higher speeds. It has been said that the best use given the UDLX was in transporting M-M field men around to dealers. Needless to say, the UDLX attracted a lot of attention then as now.

In the end, only about 150 of these stylish, but not really practical, tractors were built. Today they are one of the most admired collectible vintage tractors of all time.

Minneapolis-Moline UDLX at work, 1930s
A Comfortractor pulls a Minneapolis-Moline corn harvester through a bumper crop.

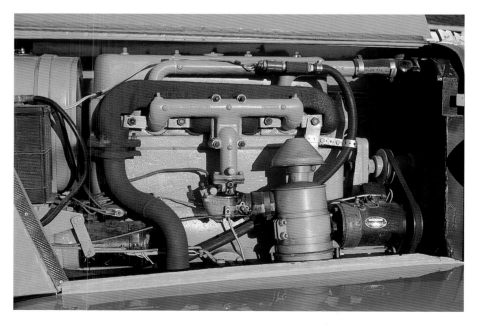

1938 Minneapolis-Moline UDLX

One of the first tractors to have a fully enclosed factory cab, the UDLX was designed to allow a farmer to work all day and then drive the tractor to town in the evening. It had seating for three, wind-up windows, windshield wipers, a speedometer with an odometer, cigarette lighter, and a shift-on-the-fly five-speed transmission. Owner: the Keller family.

Hydraulic Power: A Revolution in Utility

Probably the greatest step forward for tractor utility came from the addition of hydraulic power. Today, with hydraulic power as common as dust on a windowsill, it is hard to imagine why it was so long in coming. There are two reasons: High-pressure pumps were not available, and high-pressure cylinders did not have adequate life in dirty conditions.

The first tractor hydraulics appeared in 1934 as an implement lift on the John Deere Model A. It was fairly low pressure and was enclosed in the final-drive housing to keep it out of the dirt. Other manufacturers followed with similar systems, culminating in the famous Ford-Ferguson three-point hitch of 1939.

Ferguson's patented system was different from Deere's in that it had a feedback control system. Control-lever movement created an error signal in the valve pack, and the hydraulics moved the implement to correct the error. The secret of Ferguson's system was that implement overload could also create the error signal. Thus, if the plow hit harder soil, it would raise automatically until the draft load was the same as had previously been set by the operator. This raising and lowering to maintain the set draft load happened instantly without operator input. It made tractor plowing a job even the inexperienced could do. To prove the point, Henry Ford had an eight-year-old boy demonstrate plowing with his new 9N at the press introduction in 1939.

1959 Ford 961 Diesel

The 901 Series tractors were 50-hp row-crop machines offered by Ford from 1958 through 1961. The 961 had a live PTO and five-speed transmission. It used the 172-ci (2,817-cc), four-cylinder engine of the 801 Series. Owner Floyd Dominique is at the wheel.

1959 Case 300B Tripl-Range

Above, both photos: *The 300B replaced the Rock Island–built 300, which in turn had replaced the VA. Both the 300 and 300B used a Case 148-ci (2,424-cc) gasoline engine or a Continental 157-ci (2,572-cc) diesel. Tripl-Range drive consisted of a four-speed transmission plus a three-speed auxiliary, for twelve forward speeds and three in reverse.*

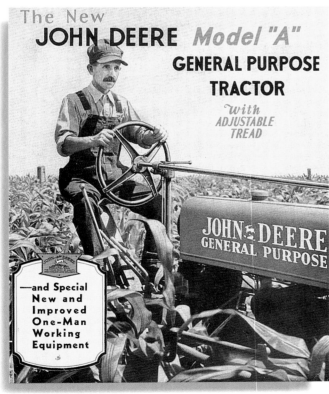

1930s John Deere Model A brochure
The Model A was first offered in 1934 and was the first tractor with a hydraulic implement lift.

1945 Allis-Chalmers Model C
Above, both photos: *This Model C has a Northfield loader conversion. The C was essentially the same as the Model B, but with a tricycle front end. The C was offered from 1939 to 1949. It was followed by a Model CA, built from 1949 to 1958 with a more powerful engine and adjustable wheel spacing. Owner: Kenneth Anderson.*

1952 John Deere Model AWH
Although 1952 was the last year for the Model A general-purpose line, some AO Orchard and AR standard-tread versions were built in 1953. This rare wide-front Hi-Crop was one of the last of its type. The A was at first unstyled, then in 1938 it received styled sheet metal. In 1947, it was restyled into what is known as the "late-styled" configuration. Owner: Larry Maasdam.

Ford-Ferguson 9N

Ford Motor Company of Dearborn, Michigan

The Ford-Ferguson was one of the greatest and most significant tractors of the twentieth century. It heralded the change from the row-crop configuration to the now-standard utility configuration, and introduced the load-compensating three-point hitch. Although small—weighing just 2,300 pounds (1,035 kg) with a 120-ci (1,966-cc) engine—it could plow 12 acres (4.8 hectares) a day with two 14-inch (35-cm) plows. This was a plowing rate that only a much larger and more expensive John Deere Model G or Farmall Model M could equal. More than 840,000 Ford N Series were sold over twelve years at a price starting at less than $600, eventually raising to about $1,200. At the peak of their popularity, 9,000 tractors were being delivered per month.

The Ford-Ferguson 9N was introduced in 1939; the "9" in the model designation stood for the year. Irishman Harry Ferguson had given up trying to build tractors with gear-maker David Brown in England and had made a deal with auto magnate Henry Ford. Ferguson had been working on his system since the 1920s, first applying it to the original Fordson tractor, then to a tractor of his own design, the Ferguson-Brown, and finally, the highly successful Ford-Ferguson. In 1942, a 2N wartime version was introduced, which was

Ford N Series engineer Harold L. Brock, 1990s

Harold L. Brock displays Harry Ferguson's spring-wound tractor model that showed the basic workings of the three-point hitch system. This model was first shown by Ferguson to Henry Ford at the meeting that formed the famous Handshake Agreement. (Photograph © Robert N. Pripps)

1941 Ford-Ferguson 9N

The 1941 version of the famous 9N can be identified by the liberal use of chrome, such as on the radiator cap and shift knob. Owner: Jack Crane.

1939 Ford-Ferguson 9N

This 9N was probably built during the first week of production and still sports its cast-aluminum hood. Because of their propensity for damage, such hoods are rare today. Owners, such as Palmer Fossum who owns this one (serial number 364), often polish them to shine like chrome.

essentially the same as the 9N, but was originally built without an electrical system or rubber tires. These items later found their way back onto the 2N, but the designation stayed the same, since it allowed a wartime price increase as well.

In 1948, the modernized 8N arrived, but Ferguson was excluded from the deal by Ford Motor Company's new president, Henry Ford II. Young Henry, grandson of the auto magnate, discovered that the company had already lost $25 million selling tractors to Ferguson, who re-sold them through his tractor and implement dealerships. Young Henry's interpretation of the famous Handshake Agreement between his grandfather and Ferguson was that either party

1930s Ford-Ferguson 9N brochure

could break it at any time without cause. While Ferguson agreed with that part of it, he didn't agree that the 8N could incorporate his patented Ferguson System. A rancorous and expensive lawsuit followed.

In the end, Ford settled by paying Ferguson $12 million—a fraction of the original suit—and by producing the all-new NAA Jubilee tractor, which had a different hydraulic system. The basic patents on Ferguson's three-point hitch had run out by then, and all manufacturers were jumping aboard. Ferguson went on with his own version of the tractor, the TE-20 built in England and the TO-20 built in Detroit. Ferguson later sold out to Massey-Harris, which then became Massey-Ferguson.

1945 Ford-Ferguson 2N
This 2N has a two-bottom disk plow mounted on the three-point hitch. The 2N used a 120-ci (1,966-cc), L-head, four-cylinder engine, which gave a belt rating of 23 hp. Owner: Dean Simmons, Frederickstown, Ohio.

1946 Ford-Ferguson 2N
Built from 1942 through most of 1947, the 2N began life as a wartime version without starter, generator, or rubber tires. Later, as these items were removed from the list of critical war effort parts, they found their way back into production. 2Ns were painted a solid deep gray at the factory. This one, owned by Doug Marcum, is painted in 8N colors.

1952 Ford 8N and 1960 Ford 601 Workmaster

These two tractors have been "zero-timed" by N-Complete of Wilkinson, Indiana. More than a restoration, they have been remanufactured to completely new standards and carry a new-tractor warranty.

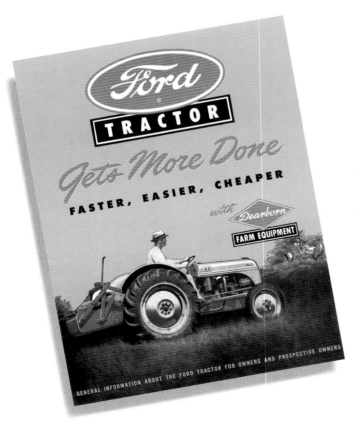

1952 Ford Model 8N

Above: *This was the final version of the venerable 8N, the largest-selling Ford tractor after the Fordson Model F—more than 520,000 8Ns and 750,000 Fordson Fs were built. Collector Palmer Fossum is at the controls.*

Left: 1950s Ford 8N brochure

THE HARRY FERGUSON GENIUS

"You haven't got enough money to buy my patents."
—Harry Ferguson to Henry Ford at the time of the Handshake Agreement, 1938

Henry George "Harry" Ferguson was born in Growell, Northern Ireland, in 1884. He was the fourth of eleven children, which was more than the family farm could support, so he followed his older brother into town and into the budding automobile business. To promote his brother's garage, Harry became a race-car driver, chalking up several victories. He next turned to airplanes, and in 1909, just six years after the Wright brothers flew, Ferguson took to the air in a monoplane of his own design. In fact, he won a £100 prize for being the first to make a three-mile (4.8-kmh) flight in the Irish town of Newcastle.

Racing and flying had made Ferguson quite famous. With his fame and winnings, he opened his own automobile business. He then hired twenty-year-old Willie Sands as a mechanic. Sands, as it turned out, had a natural aptitude for engineering. He would continue to be Ferguson's right-hand man into the 1950s, and his contributions to Ferguson's success were immense.

In 1914, Ferguson became a dealer for the Overtime tractor, the overseas version of the Waterloo Boy. When World War I broke out, the British government made every effort to gather all available tractors to produce food. When domestic tractors were not available in sufficient quantities, the government acquired all available imports from the United States. Ferguson was employed to keep records of government-owned tractors in Northern Ireland. He and Sands traveled the country, gaining experience with the good and bad features of several brands.

There were by that time a considerable number of Ford Model T cars in Britain, and several companies were offering kits to convert the Ford into a tractor. Ferguson and Sands devised a plow for one of these kits in which the draft load was reflected to the underside of the Ford ahead of the rear wheels. Increased draft loads pulled down on all four wheels, increasing traction and eliminating any tendency to rearing.

By 1918, there were some 6,000 of Ford's own new lightweight tractor plowing in England. Before the end of the year, the first government driver had been killed in a backflip accident. The tractor, subsequently to be called the Fordson, was notorious for this fault. When the plow struck an obstacle, the lightweight tractor would rotate around its own rear axle, pinning the hapless driver to the ground. Ferguson, Sands, and their team—which now included Archie Greer and John Williams—applied their experience with the Model T plow to the Fordson.

The result was a plow called the Duplex Hitch Plow. It was semi-rigid in the vertical plane, preventing back-flips, but free in the lateral plane, to allow steering. To counter the tendency for the plow to come out of the ground when the front wheel dropped into a hole, an ingenious linkage system to a skid at the rear made the plow move counter to the front of the tractor. This patented feature was called a floating skid. The team then harnessed the new field of tractor hydraulics to the Duplex Hitch Plow.

When U.S. Fordson production was discontinued and production was slow to begin in Ireland, Ferguson's team bought components of their own design and made a tractor known simply as the Black Tractor, because of its paint color. It was about 8/10-scale version of a Fordson, but it had built-in hydraulics and used a three-point hitch. Ferguson demonstrated the tractor around the country and won many accolades. He finally persuaded David Brown to join him in building the tractor, which was then known as the Ferguson-Brown.

Sadly, the Ferguson-Brown did not sell well. Brown would not build in sufficient quantities to get the price down, and the little tractor would not sell at the price required to cover the costs of small lots. It was at that time that Ferguson and Brown parted ways, and Ferguson joined with auto magnate Henry Ford to make the Ford-Ferguson 9N.

1933 Ferguson-Brown "Black Tractor"
The "Black Tractor" created by Harry Ferguson and British gear maker David Brown bore a suspicious resemblance to the Fordson. It used a Hercules engine with a David Brown transmission and differential.

1955 Ferguson TED-20
The TED-20 was the version of the basic English-built "Fergie" designed for distillate, or Tractor Vaporizing Oil (TVO).

While Ferguson had undeniable mechanical talent, he was primarily a feisty promoter. His enthusiasm for the task was so infectious that he was able to surround himself with all the other talents he needed. With shameless chutzpah, he talked Ford into advancing him $50,000 to get started. When the Ford-Ferguson deal dissolved in 1947, Ferguson was a millionaire, and Ford had lost $25 million in selling tractors to Ferguson.

In retrospect, there is disagreement as to whether Ferguson should get the lion's share of the credit for the 9N and the three-point hitch. Engineers involved with the project also heap praise on Sands, Williams, and Greer—as well as the superb engineering and manufacturing capability of the Ford Motor Company and its N Series design engineer Harold L. Brock. In any case, it was Ferguson who pulled it off.

It was also Ferguson who built his own Ferguson version of the N Series tractor on both sides of the Atlantic when the famous Handshake Agreement was dissolved—and he out-sold Ford at Ford's own game. It was Ferguson who marketed a line of extremely clever implements for the three-point hitch. And it was Ferguson who caused his name to grace the finest farm machinery, both as Ford-Ferguson and Massey-Ferguson, to this day.

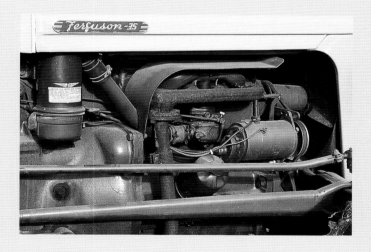

1955 Ferguson TO-35
A 134-ci (2,195-cc), four-cylinder, OHV Continental engine powered the TO-35.

Allis-Chalmers Model B
Allis-Chalmers Company of Milwaukee, Wisconsin

1938 Allis-Chalmers Model B

The B was introduced in late 1937 as a 1938 model. It weighed just a ton (900 kg) and originally cost $495. Until mid-1940, standard equipment did not include an electrical system. A variety of custom implements were offered, making the B a handy tractor for the small farm. Drawbar power was about 10 hp.

In response to the success of the Farmall, Allis-Chalmers introduced its small, inexpensive Model B in 1938. It was a one-plow machine weighing less than one ton (90 kg) and featuring a wide front arched axle.

It was designed as a replacement for a team of horses, which still provided most of the motive power for pre–World War II farms. Front and rear tread was adjustable by reversing wheels and by changing wheel clamps; a fully adjustable front axle was available later. With rubber tires and an electrical system, the Allis B, as it was affectionately called, sold for less than $600.

1939 Allis-Chalmers Model B
Above left: *A Model B mounted with a one-row cultivator. Owner: Keith McCaffree.*

1951 Allis-Chalmers Model IB
Above right: *The IB was a low, compact version of the B built for industrial uses. It had foot controls like a car—including a gas pedal—for easier operation by people not used to tractors. Owner: Gaylord De Jong.*

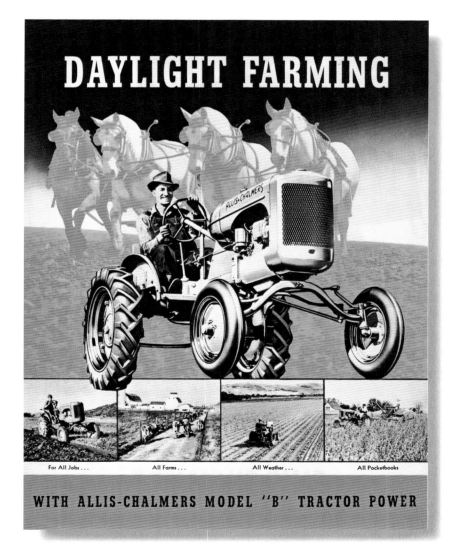

1940s Allis-Chalmers Model B brochure

Orchard Tractors: Streamlined Form and Function

In the late 1920s, modified versions of standard tractor models were first offered for specialized agricultural chores. The innovative John Deere offered some of the first of these machines, which included high-clearance tractors and wide and narrow front ends. Deere soon released a streamlined orchard machine, with rear fenders that enclosed the wheels and other cowling designed to aid the tractor in slipping through the low branches of orchard without harming the valuable trees.

1937 McCormick-Deering O-12
The O-12 was the orchard version of the standard-tread W-12. The row-crop version was much more widely known—the F-12 Farmall. Only 4,000 O-12 machines were built between 1932 and 1940. Owner: Dan Schmidt.

1947 Case VAO

Above, all photos: *The orchard version of the popular Case VA Series, the VAO was built from 1942 to 1955. The engine was a 124-ci (2,031-cc), OHV four, giving about 20 belt hp on gasoline; a distillate manifold was an extra-cost option. The VAO weighed about 2,700 pounds (1,215 kg). The Eagle Hitch and hydraulics came out on the VAO in 1949, but older tractors could be retrofitted. Owner: Larry Maasdam.*

1950s Oliver 70 Orchard brochure

The Fuel Revolution: From Diesel to "Greased Air"

Kerosene was the undisputed tractor fuel of choice until the late 1930s. It was cheap, stored well, and generally provided satisfactory operation. The quality of gasoline varied greatly, and engine designers had to engineer for the lowest expected gasoline grade, losing much of gasoline's power advantage over kerosene. Still, a small quantity of gasoline was required for kerosene-engine starting, and engines typically needed to be shut off on gasoline as well, or the kerosene in the carburetor would make starting difficult next time.

With the advent of octane ratings and standards for gasoline in the mid-1930s, gasoline gained in popularity. Gasoline's benefits came in large part from higher compression, so the same size engine could generate more horsepower. Side benefits came from easier starting, easier vaporization, and lower fuel consumption.

Gasoline was still in the process of supplanting kerosene and its variations such as distillate, or tractor vaporizing oil (TVO) when the diesel engine arrived on the scene. Caterpillar was first with a diesel tractor, the Diesel Sixty of 1931. International Harvester followed with the first diesel wheeled tractor, the WD-40 of 1934.

The diesel engine depended on a variable-displacement, high-pressure injector pump. Making such a pump to survive the low-lubricity diesel fuel, which was a close relative to kerosene, was a difficult problem that had to be solved before the diesel was ready for everyday use. Also, the diesel's 16:1 compression ratio made the engine extremely large and heavy until better steel castings became available.

The diesel's high compression ratio and restriction-free inlet contributed to low fuel consumption per horsepower hour. Another big economy factor was the fact that diesels run on the lean side of stoichiometric combustion,

1955 Ferguson TEF-20
Ferguson's TEF-20 was similar in size and shape to the Ford 9N/2N, as Ferguson used Ford tooling in setting up production in Coventry, England. The 20 had an indirect-injection diesel engine from the Standard Motor Company, a British car maker. The four-cylinder engine developed 26 hp.

rather than on the rich side as do gasoline and kerosene engines. Diesel entirely supplanted other tractor fuels by the late 1970s.

Another fuel revolution was hailed in the 1940s with liquefied petroleum gas (LPG)—nicknamed "greased air" in the early days. Minneapolis-Moline first equipped tractors to run on LPG in 1941, but it was not until 1949 that the Minneapolis-Moline U LPG was the first tractor tested on LPG at the University of Nebraska. LPG gained popularity slowly before reaching its peak of use.

When the engine compression ratio is raised to about 9:1, LPG provides more horsepower than the same gasoline engine at 6:1 compression. Because of its clean burning, LPG engines ran better without tune-ups for longer periods of time, and oil stayed cleaner and did not need to be changed as often. The cost of operation was also less than for an equivalent gasoline engine. On the down side, LPG was more difficult to store and refuel than gasoline or diesel fuel. Ultimately, its economy was not as great as that of diesel.

Popularity of LPG peaked in the 1960s and then faded. The University of Nebraska's test in June 1968 of a Minneapolis-Moline G-900 (LPG) was the last LPG-fueled tractor test.

1957 Massey-Harris 333 Diesel

The Massey 333 was the final version of the line that started in 1939 with the Model 81. Beginning with the Model 30 of 1946, the line was considered to be capable of pulling a three-bottom plow in most soils. This is one of only 150 made between 1956 and 1958. Power came from a 201-ci (3,292-cc) Continental. A five-speed transmission with high-low range and a three-point hitch were provided. Weight was about 6,000 pounds (2,700 kg). Owner: Ken Peterman.

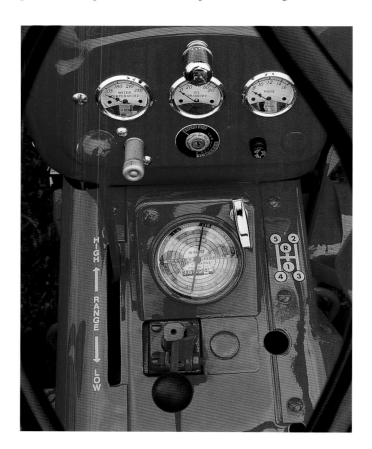

1950s Massey-Harris 44 Diesel Standard

Both photos: *The 44 was possibly the best tractor ever built by Massey-Harris. It was available with a Massey four, Continental six, or Massey diesel four. All produced about 46 hp, giving the tractor a three-plow rating. Owner: Dan Peterman.*

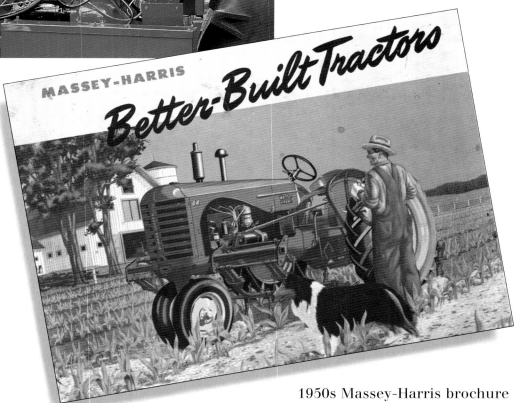

1950s Massey-Harris brochure

1954 Minneapolis-Moline GBD

Above and right: *The GB-Diesel had a 426-ci (6,978-cc) Minneapolis-Moline diesel with Lanova-type combustion chambers and produced 63 hp. The tractor weighed about 8,200 pounds (3,690 kg). Owner: Keith and Adam Bruder.*

1950s Massey-Harris 44 Vineyard Special

Left: *The Vineyard version of Massey's famed 44. Owner: Larry Maasdam.*

1960 John Deere 830 Rice Special

The 830 was the last of the big two-cylinders from Deere. It featured a 472-ci (7,731-cc) diesel turning a rated 1,125 rpm and putting out 75 hp through a six-speed transmission. Weight was a little over 8,000 pounds (3,600 kg) without ballast. The 830 was rated for a six-bottom plow or a 20-ft (6-m) disk. The Rice Special version had a wider axle with mud shields and mud covers on the brakes. This one also has the 24-volt electric-start option. Owner: Ken Peterman.

1958 Farmall 450 Diesel

The Farmall 450 was built from 1956 to 1958 in regular and high-clearance as well as LPG, diesel, and gasoline versions. The 281-ci (4,603-cc) engine gave the 450 four-bottom capability. Owner: Norm Seveik.

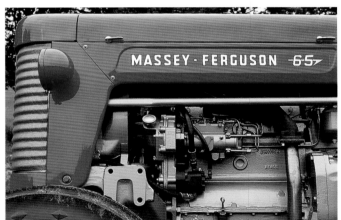

1959 Massey-Ferguson 65 Mark 1

The 1958–1965 British 65 was basically the U.S. 40 with a 50-hp Perkins indirect-injection diesel. From 1961 on, a direct-injection diesel was used. Final-drive gear reduction was made within the hubs to compensate for the larger tires required by the increased power. Owner: Mike Thorne.

1960 John Deere 520 LPG

Left and below: *The 520 LPG had a two-cylinder engine rated at 38 PTO hp. Owner: Melanie Maasdam.*

1960 Oliver XO-121

Left: *The XO-121 was an experimental tractor powered by a 199-ci (3,260-cc) engine with a 12:1 compression ratio for high-octane gasoline. The engine produced 57.5 brake hp at a specific fuel consumption of 0.385 pounds per hp hour—a rate comparable to diesel engines. The tractor now resides in the Floyd County Historical Society Museum in Charles City, Iowa.*

Oliver Hart-Parr 70

Oliver Farm Equipment Corporation of Chicago

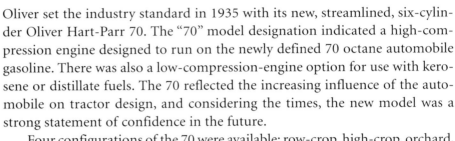

Oliver set the industry standard in 1935 with its new, streamlined, six-cylinder Oliver Hart-Parr 70. The "70" model designation indicated a high-compression engine designed to run on the newly defined 70 octane automobile gasoline. There was also a low-compression-engine option for use with kerosene or distillate fuels. The 70 reflected the increasing influence of the automobile on tractor design, and considering the times, the new model was a strong statement of confidence in the future.

Four configurations of the 70 were available: row-crop, high-crop, orchard, and standard-tread. Optional equipment included a starter and lights. The tractor was marketed in Canada by the Cockshutt Farm Equipment Company of Brantford, Ontario, as its Cockshutt 70.

In 1937, the Hart-Parr name was dropped completely, and the 70, along with other Oliver/Cockshutt tractors were restyled. Under the new Fleetline styling, the 70 remained much the same as before. Rubber tires and a six-speed transmission became standard equipment in 1939.

1935 Oliver Hart-Parr 70

With its six-cylinder engine, the Oliver 70 of 1935 was a remarkable leap forward in tractor technology. It was one of the first to use stylish sheet metal to lure buyers. With a self-starter and convenient hand controls, it was also one of the first to cater to the needs of women and young people as tractor operators. The tractor was also one of the first designed to run on the new 70-octane gasoline, hence the designation "70." A low-compression version was available for those not ready to give up their low-cost kerosene fuel.

Orphan Tractors: The Second Industry Debacle

The tractor makers who had survived the economic recessions and depressions leading up to World War II, looked forward to the postwar years as a new, golden dawn when power farming would make them rich. These newfound halcyon days were short lived, however. In 1954, yet another economic recession struck farm-equipment manufacturers, and more makers fell by the wayside. In 1930, there had been some forty-seven tractor builders in North America. By 1960, there were but twelve.

1937–1938 Co-op Duplex No. 2
The Duplex No. 2 was built by the Co-operative Manufacturing Company of Battle Creek, Michigan. It was powered by a 201-ci (3,292-cc), six-cylinder Chrysler engine and used a four-speed transmission. Owner: Don Wolf.

1940–1941 Co-op B2JR
Built by the Arthurdale Farm Equipment Corporation of West Virginia, the B2JR was built from 1940 through 1941. It featured a 201-ci (3,292-cc), Chrysler industrial engine and four-speed transmission. Governed to 1,500 rpm, the engine produced 33 hp. Owner: Larry Maasdam.

1936 Eagle 6B

The Eagle Manufacturing Company of Appleton, Wisconsin, made a long line of innovative tractors beginning in 1906. In 1930, the modern Eagle 6A standard-tread was introduced using a 358-ci (5,864-cc), six-cylinder Hercules motor. It was followed by the 6B row-crop in 1936 with a smaller Hercules six. A 6C that debuted in 1938 was the standard-tread version of the 6B. Production ended in 1940. Owner: Richard Grimm.

1938 Avery Ro-Trak

Made in Peoria, Illinois, by the Avery Farm Equipment Company, the Ro-Trak was convertible from a narrow to a wide front configuration. Since there was no axle pivot as on conventional tractor front ends, the vertical tubes contained soft springs to allow the front wheels to go over bumps without undue frame twist. An unusual feature was that when the brakes were applied, the front dipped like a soft-sprung car. A 212-ci (3,473-cc), L-head, six-cylinder Hercules engine was used. No power rating was given, but the Ro-Trak was called a two-to-three-plow tractor. It had a three-speed transmission giving a road speed of 16 mph (25 kmh). Production ended in 1941.

High-Clearance Tractors: Standing Tall

The row-crop tractor was born to allow cultivation of tall crops, such as corn. It soon became evident, however, that the normal row-crop machine still did not provide adequate clearance for late-growth tilling. To meet this new demand, most manufacturers developed high-clearance versions of their row-crop tractors. These only sold in limited numbers and are thus now among the most sought after of tractors by collectors.

1956 Case 400 HC

Right: *This Case 400 High-Clearance is one of only 150 made. It had the Eagle Hitch. The Case 400 was available in gasoline or diesel versions; this one is gasoline. Owner: Jay Foxworthy. This 400 HC came from the sugar cane fields of Louisiana.*

1948 Case VAH

The high-clearance member of the Case VA Series, the VAH was built from 1948 to 1955. The engine was a 124-ci (2,031-cc), OHV four, giving about 20 belt hp on gasoline; a distillate manifold was an extra-cost option. The Eagle Hitch and hydraulics were available for the VAH in 1949. Owner: Norm Seveik.

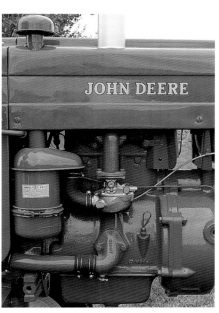

1955 John Deere 40V

Only 310 of the rare 40V were built, all in 1955. Higher than the standard 40, the 40V gave 26 inches (65 cm) of crop clearance. The 40 had a vertical two-cylinder engine of 100.5 ci (1,646 cc), which gave it 25 belt hp. Owners: Ken and Dan Peterman.

1950s Massey-Harris 44 Diesel Special High-Crop

Below, both photos: *Only two examples of the Massey 44 Diesel High-Crop are known to exist. The engine was built by Massey-Harris and rated 45 hp. Owner: Larry Maasdam.*

1955 Farmall 400 HC

Above and right: *The 400 was the successor to the mighty Farmall M and Super M, but was only built in 1955 and 1956. It was then replaced by the 450. This rare 400 High-Clearance was equipped with the Torque-Amplifier dual-range power-shift gearbox. Owner: the Keller family.*

The Evolution of Live PTO and Hydraulics

"The tractor had been a mechanical substitute for Old Dobbin, and ate gas instead of hay. But the general-purpose tractor, with its power take-off, planted, cultivated and harvested. It practically wiped out the 'doubtful' and 'non-tractor' operations."
—Prairie Farmer, 1941

1940s Case Model SC

The Case Model SC was announced in November 1940 and was all-new from top to bottom. The "plain" S was the standard-tread version; there were also Orchard (SO), Industrial (SI), and the SC-4 versions. The SC-4 was only built in 1953–1954 and had a fixed-tread wide front axle.

Tractors had power takeoffs in the form of belt pulleys from the earliest days of steam. In the true sense of the word, however, the term "power takeoff" implied an output for supplying tractor engine power to an implement while the tractor is in motion, which would not apply to a belt-pulley drive. The McCormick-Deering 15/30 of 1921 was the first tractor to be equipped with a true PTO. This PTO, and most others until the mid-1950s, was driven by gears downstream of the clutch; when the clutch was disengaged, the PTO no longer received power from the engine.

There were several early examples of "live" PTOs, or PTOs that either had their own clutch or were continuously connected to the engine. An example of the former is the 1924 Hart-Parr 12/24 E: It had an optional PTO driven from the clutch end of the engine by means of a separate clutch. A series of shafts and universal joints carried the power back across the operator's platform, terminating by the drawbar. Another example of the live PTO is the cable winch drives used by Caterpillar and others to control trailed earthmover scrapers.

The traditional live PTO as known today with a two-stage clutch pedal (one stage disconnects the engine from the transmission, the second stage disconnects the PTO) first debuted on the Cockshutt Model 30 of 1946.

The situation with hydraulic power was similar. Initially, pumps were mounted near the transmission/differential and driven by gears downstream of the clutch. Often, the same powertrain was used for the pump as for the PTO.

Later, as efficient high-speed hydraulic pumps became available, pumps were driven directly by the engine, thereby allowing for "live" hydraulics. Deere offered an optional live hydraulic pump on its 1948 Model B. It was not until the early 1950s that others followed suit.

1930s Cockshutt Model 30

The Cockshutt Plow Company of Brantford, Ontario, dated back to 1877. For years, it marketed tractors made by others, mostly Oliver, with the firm's own colors. In 1946, however, Cockshutt released the first tractor of its own design, the Model 30. It was the first production tractor to have a live PTO of the type used today. The engine was a 30-hp, four-cylinder Buda.

Farmall MD

International Harvester Company of Chicago

Beginning with the 1924 Regular, IHC's McCormick-Deering Farmalls were the first volume-production all-purpose tractors. Representing the top of the line, the Model M, was launched in 1939. The big, three-plow M was also the best-selling Farmall. It later evolved into the Super M-TA with the first power-shift underdrive. It was available in engine configurations for distillate, gasoline, and LPG fuels, but the diesel version, which came out in 1941, revolutionized farm power and normalized the use of the diesel engine.

The first diesel tractor was the Caterpillar Diesel Sixty of 1931, followed in 1934 by McCormick-Deering's WD-40. The WD never gained the popularity of the MD, however. Sales of the MD averaged 22,000 units per year, even though the MD cost 50 percent more than a gasoline version. Fuel consumption was about one-third to one-half that of the gasoline engine.

McCormick-Deering diesels had a unique starting system. The engine was equipped with a spark-ignition setup as well as the diesel-fuel system. A compression-release lever enlarged the combustion chamber, diverted intake air through the carburetor and activated the carburetor float. Once running on gasoline, the compression lever was thrown, which also engaged the diesel injectors. The engine was then running on diesel.

The MD was a big tractor for its time, weighing in at 5,300 pounds (2,385 kg) in 1941 and growing to

1941 Farmall MD

A diesel-powered version of the great Farmall M debuted in 1941 as the MD and won many converts. The MD was upgraded with M-W aftermarket auxiliary gearbox, power steering, live hydraulics, and live PTO. Owner: Alan Smith.

5,900 pounds (2,655 kg) by 1952. An additional 3,000 pounds (1,350 kg) of ballast was not unusual. Originally, the four-cylinder, 248-ci (4,062-cc) engine produced a maximum of 35 hp on the belt pulley. By the end of production, internal improvements resulted in the availability of almost 40 belt hp.

A Super MD debuted in 1952 with a four-cylinder, 264-ci (4,324-cc) engine and about a third more power. Weight was up to 6,000 pounds (2,700 kg), without ballast. Farmers claimed a 2.5-acre (1-hectare) per hour plowing rate for the Super MD with fuel consumption of a gallon (3.3 liters) per hour.

1954 Farmall Super MD-TA

The MD-TA was the diesel version of the famous M with the lever-controlled Torque Amplifier. Super M Farmalls were built from 1952 to 1954. The 264-ci (4,324-cc) engine gave the Super a 32 percent power boost over the previous MD. In addition to the MD, gas and LPG versions of the Super M, with and without the Torque Amplifier, were also available. Owner: Donald Schaeffer.

Allis-Chalmers Model WC and WF

Allis-Chalmers Company of Milwaukee, Wisconsin

Allis-Chalmers introduced its two-plow WC in 1933, the first tractor to be offered with rubber tires as standard equipment; steel wheels were optional. A channel frame was employed, which was lighter and less expensive than the castings used on the Model U, and the WC weighed only 3,200 pounds (1,440 kg). The 201-ci (3,292-cc) engine featured a 4.00x4.00-inch (100x100-mm) bore and stroke, and was rated at 1,300 rpm. Both kerosene and gasoline versions were available. The WC was a row-crop tractor; its standard-tread running mate, the WF, was introduced in 1940 and was produced through 1951.

Allis-Chalmers built more than 186,000 WC and WF tractors.

The popular WC was succeeded by the WD in 1948. The WD boasted the first power-adjustable rear wheel tread and was also one of the first tractors available with a live power takeoff (PTO). The same engine was used as in the WC, but its rated speed was upped to 1,400 rpm. The WD was available in dual tricycle, single, and adjustable wide front ends.

The WD was followed in 1953 by the WD-45, built along the same lines. The engine of the WD-45 had a 4.50-inch (112.5-mm) stroke, giving it substantially more power. It could be had in gasoline, dual-fuel, or LPG versions. In 1955, the WD-45D six-cylinder diesel was offered.

1941 Allis-Chalmers WF

The WF was the standard-tread version of the famed WC row-crop and was built from 1940 to 1951. They were both two-plow tractors and featured the first "square" engine in the industry, with bore and stroke being equal at 4 inches (100 mm).

1940s Allis-Chalmers WC brochure

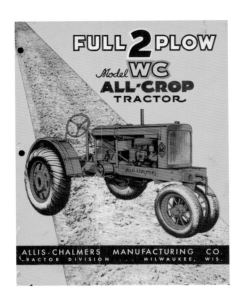

1930s Allis-Chalmers WC brochure

Above: *The two-plow WC was introduced in 1933. It was the first tractor to be offered with rubber tires as standard equipment; steel wheels were optional. A channel frame was employed, and the WC weighed only 3,200 pounds (1,440 kg). The 201-ci (3,292-cc) engine was rated at 1,300 rpm.*

1947 Allis-Chalmers WF

Right, both photos: *The WF featured a four-speed transmission and 201-ci (3,292-cc), four-cylinder, OHV engine of 28 hp. It sold for about $1,200 in 1947. Weight was about 3,500 pounds (1,575 kg). Owner: Paul Mihalovich.*

190

Tractors of Many Colors

From the 1930s through the 1950s, tractor makers began to settle on distinctive colors and paint schemes to set their machines apart from the competition. These trademark colors became famous in the industry—and tractors of a brand's "color" became easily identifiable in the field.

1946 Farmall H
The Farmall H was introduced in 1939 and used a 152-ci (2,490-cc), OHV four-cylinder engine. Most were equipped for gasoline, but a distillate option was available. Owner Leon E. Geiss bought this H new in May 1946. It still has the original wartime S-3 tires.

1948 Farmall MV

Above, both photos: *This unusual Farmall high-clearance model was ordered with a single front wheel. The MV could be had with either the wide front end or the single front wheel. The rear wheels were driven by drop boxes at the end of each axle, which raised the axle to provide more clearance. Owner: Norm Seveik.*

1948 Farmall HV

The HV was some 10 inches (25 cm) taller than the standard H, allowing the cultivation of taller crops such as corn and sugar cane. IHC made a cotton picker based on the HV, and some have been converted to the much-sought-after HV. This came from the factory as a single-front-wheel HV. Owner: Larry Maasdam.

1948 Oliver 60 Standard

Above and left: *To compete with small tractors from the other makers, Oliver offered its 60 in 1940, and production continued through part of 1948. The 60 was a scaled-down 70, offering a four-cylinder engine rather than the six. Belt power was about 18 hp. Gasoline and distillate fuels were options, as were row-crop or standard-tread versions.*

1950s Oliver advertising poster

1950s Oliver Super 55

Above: *Built from 1954 to 1958, the Super 55 was Oliver's answer to the Ford 9N, 2N, 8N, and NAA tractors with three-point hitch. Of the same basic configuration as the Ford N-Series, the Super 55 didn't leave anything to chance. Oliver offered a six-speed transmission (instead of three or four), gasoline or diesel engines (Ford only offered gasoline), and a 144-ci (2,359-cc) engine (Ford's largest was the 134-ci/2,195-cc in the NAA). Oliver may have thought it had roused the sleeping giant, however: In 1955, Ford's models began to proliferate in all directions.*

1950s Oliver 77 brochure

This cutaway drawing showed the heart of the 77: Oliver's six-cylinder powerplant.

1959–1960 Oliver 440

Built in 1959 and 1960, the 440 was the culmination of the line that began with the Super 44 of 1957–1958. The two models were the same, except for the paint. The 2,000-pound (900-kg) tractor used a 20-hp Continental engine. The offset configuration, with the engine set to the left and the steering wheel and driver's seat to the right, ostensibly enhanced the visibility when using an under-belly cultivator.

1960s Oliver 660

Introduced in 1959, the 660 was an improved and restyled Super 66. Gasoline or diesel engines of 155 ci (2,539 cc) were offered. Disk brakes were standard, but power steering was an option. Production continued through 1964.

1960s Oliver 880

Above and right: *The 880 was the last of the line that began with the 18/27 of 1930. It went through the 80, 88, and Super 88 before reaching its final version. The 880 was built from 1958 to 1963. It featured improvements such as a power-shift torque amplifier, standard power steering, and load-biased three-point hitch. Gasoline, diesel, or LPG, the 880 was in the 60-hp class.*

1950 Case Model D

The three-plow Model D replaced the Case C and was basically a styled version of its predecessor. It was the best-selling Flambeau Red tractor with around 100,000 sold from 1939 to 1953.

1958 Case 400B Case-O-Matic

Except for the Case-O-Matic torque converter drive, the 400B was much the same as the old 300C. The tractor featured a 148-ci (2,424-cc), four-cylinder engine and sixteen gear ratios plus the torque converter. With ballast, it weighed just over 8,000 pounds (3,600 kg). Owner: Norm Seveik.

1950 Massey-Harris brochure

1965 Case 830 CK

The 830 was a five-plow tractor with a 284-ci (4,652-cc), four-cylinder engine. A dual-range conventional transmission was available, but the Case-O-Matic torque converter drive was available. Besides gasoline, LPG and diesel versions were also offered.

1940s Case brochure

1940 Massey-Harris Super 101

Left: *Massey-Harris's flagship Super 101 was powered by a 218-ci (3,571-cc) Chrysler six. Owner: Wayne Svoboda. (Photograph © Robert N. Pripps)*

1960 Allis-Chalmers D-12

Allis-Chalmers D-10 and D-12 tractors were built simultaneously from 1960 to 1967—and were essentially the same. The D-10 was a single-row machine, while the D-12 was made for two rows and had adjustable wheel-tread spacing. Originally, an engine of 139 ci (2,277 cc) was used, but in 1961, a 149-ci (2,441-cc) engine was substituted. From 1964, the tractor had a 12-volt electrical system and a draft-load compensating three-point hitch. Owner: Helle Farm Equipment.

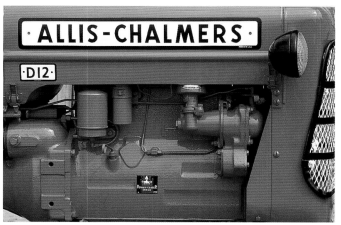

199

1953 Minneapolis-Moline BF

Above and right: *Minneapolis-Moline's BF tractor began life in 1939 as the Cletrac-General CC, made for Cletrac by B. F. Avery Company of Louisville, Kentucky. In 1951, this tractor became the M-M BF. By that time, its four-cylinder engine displaced 133 ci (2,179 cc). Owner: Vernon Parizek, president of the M-M Collector's Club.*

1941 Minneapolis-Moline ZTU

Above: *The Z Series was built from 1937 to 1956. Originally, the Z used a 186-ci (3,047-cc) engine, but by 1950, it had been increased to 206 ci (3,374 cc). Owners: Don and Betty Zwicky.*

1950 Minneapolis-Moline U

Right: *The U was powered by a 284-ci (4,652-cc), four-cylinder engine of 40 hp along with a five-speed transmission. Owner: James Taylor. Taylor's father, Russell Taylor, bought this U new. James first drove it when he was fifteen years old. He still uses it around the farm, so it isn't retired yet.*

1939 Minneapolis-Moline R

Left: *The R was the smallest tractor in the M-M line, but used the same basic engine as the Z. The displacement was reduced to 165 ci (2,703 cc), and the rated rpm was also cut, giving the R 26 hp, about 10 less hp than the Z. The R could be equipped with a Comfortractor-type cab, as on this machine.*

1958 Minneapolis-Moline 5 Star

Left and below: *The 5 Star was made from 1957 to 1961, although standard-tread gasoline and LPG versions were only made in 1958. A 283-ci (4,636-cc) engine gave it 55 hp. Owner: Walter Keller.*

Ford's World Tractor

For decades, Ford was a one-tractor company. The Fordson in its heyday and the 8N at its peak each sold more than 100,000 units in a single year in the United States. These were far and away the largest-selling models in North America. The Fordson claimed 70 percent of the tractor market; the 8N outsold the entire Deere line and was second only to the entire IHC line. In England, Ford remained loyal to its Fordson E27N Major, which was built from 1945 to 1952. But then the market—and Ford—began to change.

In 1955, the Ford Motor Company went public, and Henry II had stockholders that were not interested in the altruistic motives of Old Henry. Then, in 1956, British Ford sold more tractors than the original American firm. Rather than be further embarrassed, Detroit expanded the domestic line to two engine sizes with a variety of transmission and other options, including diesel and LPG.

But sales still slumped, so the concept of the "World Tractor" came into being in 1961. By 1964, a new line of tractors was being made in new plants in Highland Park, Michigan; Basildon, England; and Antwerp, Belgium.

The World Tractor concept was successful, and by 1966, Ford was again number two in sales, just behind Massey-Ferguson of Toronto, Ontario.

1953 Ford Jubilee Model NAA

The NAA was the first all-new Ford tractor in fourteen years and was a result of the settlement of the lawsuit with Harry Ferguson over patent infringements. Because of the settlement, an all-new hydraulic system was used. Other improvements over the 8N, which it replaced, made it heavier and more powerful. Owner: the Sparks family. This NAA had just been remanufactured by N-Complete of Wilkinson, Indiana. N-Complete owner Tom Armstrong is at the controls.

1955 Ford 800

Above, both photos: *The 800, like the later 801, used Ford's 172-ci (2,817-cc) engine. The 800 was built from 1955 through 1957. Owners: Jim and Jerriann Endries.*

1960 Ford 601 Workmaster

Left: *Built from 1958 through 1961, the Workmaster used a 134-ci (2,195-cc), four-cylinder engine.*

1958 Ford 861 Powermaster

Below, both photos: *The 861 used a 172-ci (2,817-cc), four-cylinder engine and five-speed transmission. It had live hydraulics and live PTO. Power was in the 50-hp range. LPG and diesel versions were also available, as was a ten-speed Select-O-Speed power-shift transmission.*

The End of the Reign of Johnny Popper

"A New Generation of Power"
—Deere advertising slogan, 1960

The side-by-side, transverse-mounted, two-cylinder engine was common engine architecture from the earliest times. Henry Ford's first home-built gasoline engine was a two-cylinder. Early Hart-Parrs were two-cylinders, as were most International Harvester Titans and Rumely OilPulls. In some designs the pistons moved in unison; in others, in opposition.

In the case of the famed Deere two-cylinder engine, which came to Deere with its purchase of Waterloo Boy in 1917, the pistons moved in opposite directions. Being a four-cycle engine meant that the firing sequence was not evenly spaced: It was like a double-barreled shotgun where the barrels fired sequentially, reloaded, and fired again. There was something intriguing about the exhaust note of a laboring John Deere two-cylinder, which seemed to account for much of their popularity, as well as sobriquets such as "Poppin' Johnny" and "Johnny Popper."

Variations of Deere's venerable two-cylinder engine powered the smallest tractors and the largest diesels. Originally, the configuration was popular because the close proximity of the exhaust gases to the intake manifold made a good arrangement for burning of kerosene, the cheapest fuel at the time. Also, the transverse mounting eliminated the need for bevel gears. As time went on, only Deere stuck with the two-cylinder approach. It was the focus of much of Deere's advertising, stressing that fewer parts meant a more reliable tractor.

As the 1950s waned and the power race topped 75 hp, the engineers at Deere knew the two-cylinder was at

1957 John Deere 620

The 620 arrived in 1956, the successor to the famous Deere Model A and subsequent 60. It featured the draft-control three-point hitch called Custom-Powr-Trol. A 20 percent power increase came mostly through an increase in engine rpm from 975 to 1,125. Owner: Bruce Copper.

the end of its line. It was no longer practical to increase displacement and still get the engine under the hood of a reasonably sized tractor. Further, the large-displacement two-cylinders, such as those in the Deere 820 and 830 were rated at 1,125 rpm, which was fast for an engine with an 8.00-inch (200-mm) stroke. An equivalent short-stroke, four-cylinder engine would produce the same horsepower at 2,600 rpm and half the displacement.

With secrecy rivaling the World War II Manhattan Project that created the atomic bomb, Deere management detailed selected engineers to a converted grocery-store building. Sworn to secrecy, the engineers began designing a whole new line of completely modern multi-cylinder tractors. They knew they would have just one chance to convince their loyal customers that these new three-, four-, and six-cylinder machines were worthy to bear the name of John Deere. Marketing did its part too. A gigantic hoopla was planned and carried off on August 30, 1960, at the Dallas, Texas, Coliseum. The event was known as Deere Day in Dallas. The firm flew in dealers and press people from all over the country to introduce what they called the New Generation of tractors. Besides the tractors, there were big-name entertainers, fireworks, and barbecues.

The New Generation consisted of four lines: the 30-hp 1010, 40-hp 2010, 55-hp 3010, and 80-hp 4010. Most were available in gasoline, diesel, or LPG versions and in all the configurations that were previously offered, from utility to row-crop. The 1010 was also available as a crawler.

The new tractors were nimble, well balanced, and thoroughly up to date. They were also nicely styled by famed industrial designer Henry Dreyfuss. The dealers liked them and farmers bought them. Deere had created a new generation of tractors that upheld the reputation built by the beloved Johnny Poppers.

1960 John Deere 630S

The standard-tread version of the powerful 630 was quite rare. The 630 was available with gasoline, "all-fuel," or LPG engines; a diesel was not offered. The 630 was offered by Deere from 1958 through 1960 to replace the 620 Series. It could handle a four-bottom plow with its 321-ci (5,258-cc), two-cylinder engine and six-speed transmission. The fenders and a big oval "low-tone" muffler were new on the 630. Owner: Larry Maasdam.

1959 John Deere 730

Above and left: *Still working after forty-two years, this gas 730 with power steering used a 360.5-ci (5,905-cc), two-cylinder engine rated at 1,125 rpm and 60 hp. Not restored, this 730 is in original finish. Owner: Ken Kass.*

1959 John Deere 530

The final version of the series that started with the venerable Model B, the 530 was a thoroughly modern general-purpose tractor. Owner: Kent Bates. It is shown in front of a 1926 Gordon Van Tyne kit-built barn. These kits included everything except the gravel and water for the cement.

1960 Deere New Generation brochure

Deere's new generation of three-, four-, and six-cylinder tractors officially arrived on August 30, 1960, when they were launched to dealers at the Dallas, Texas, Coliseum. It was the end of an era—and the beginning of a new one.

Classic Tractor Replicas: A Labor of Love

The restoration of a vintage tractor is daunting work, but some tractor enthusiasts choose to go even a step further. The creation of working, scale replicas of classic tractor requires dedication and skill that is often rewarded by the respect of other tractor buffs at vintage tractor shows and meets.

½-Scale Rumely OilPull

Built by Glen Braun, this replica uses a two-cylinder Continental engine. Engine rotation, however, had to be reversed. Braun also made a six-bottom Rumely plow for it using bottoms from garden tractor plows. Braun also has a full-scale Rumely 20/40 Type G built in 1923.

½-Scale John Deere 3020

Ken Peterman constructed this one radio-controlled replica. The radio even controls the clutch, steering, engine start and stop, and the three-bottom plow. It is powered by a Yanmar diesel engine and has a Farmall Cub transmission. Peterman spent 2,400 hours on the project.

½-Scale Massey-Harris 44 Diesel

Above: *Powered by a three-cylinder Dakota engine, this replica was built by Ken Peterman. Peterman also built a ½-scale Massey combine to go with it.*

½-Scale Allis-Chalmers IB

Left: *Powered by a 3½-hp Tecumseh engine, this replica has a three-speed transmission that uses gears from lawn mowers. Everything is hand built, and even the belt pulley works! Builder: Gaylord De Jong, who also has a full-size IB.*

Hot-Rodded Farm Tractors

The search for more power began as soon as a tractor got stuck in a muddy field or a larger plow was available to till new acreage. Neighborhood mechanics or farmers handy with a wrench sometimes created their own "hot-rod tractors" by swapping engines or pairing two tractors into a jury-rigged machine that could do double the work.

During World War II, Ford built some pickups and small vans with four-cylinder Ford tractor engines instead of the normal sixes and V-8s. Enterprising businessmen soon realized that the switch could be made the other way around, and 9N, 2N, 8N, and NAA Jubilees were converted to Ford six and V-8 power. Converted as such, they became the most powerful tractors in their days.

These hot-rodded machines sometimes became local legends, winning their builder renown—and often a telephone call at an inopportune time when someone was stuck in the mud. In the larger history of tractors, they served as a stepping stone toward the creation of more-powerful factory-built machines and eventually, the articulated, four-wheel-drive tractor that rules the field today.

1950s Funk-Ford 8N V-8

Below and right: *The Funk brothers, foundrymen from Coffeeville, Kansas, produced adaptation kits in the early 1950s to convert Ford tractors from the standard four-cylinder engine to the 239-ci (3,915-cc) Ford flathead V-8 or 226-ci (3,702-cc) Ford six. The six produced 182 foot-pounds of torque at 1,200 rpm; the V-8 created 187 foot-pounds at 1,600 rpm. Most Funk-Ford V-8s use the dual vertical straight exhaust pipes that gives the operator unbeatable stereo exhaust music. Less than 100 Funk-Ford V-8s were believed to have been built. Owner: Robert Meyer.*

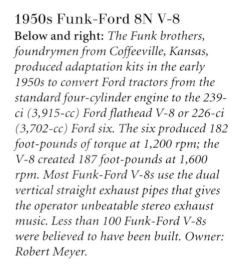

1950s Funk-Ford 8N V-8

Left: *The main reason the hoods of Funk-Ford tractors were higher than standard was due to the need for a larger radiator, and Ford radiators were supplied with the engine kits. Because exhaust gasses passed through the water jacket of the flathead V-8, it required an exceptionally large radiator. Owner: Palmer Fossum.*

1950s Farmall M V-8 Special
This Farmall M was custom built by Norm Sevick and his son Jeremy and is propelled by an International truck V-8.

1964 Fordson Power Major V-8 Special
Builder Richard Vincent mounted a 200-hp, 510-ci (8,354-cc), Perkins V-8 diesel in this Fordson Power Major in search of more plowing power.

The Development of Articulated Four-Wheel-Drive Tractors

"The Allis-Chalmers Duplex is formed by joining two 6–12s together into one unit. The transmission controls of the two tractors are interlocked so that they operate as one. . . . On many farms it is the real solution of economical tractor power for heavy plowing work."
—Johann F. "Max" Patitz, Allis-Chalmers chief consulting engineer, 1919

Articulated four-wheel-drive tractors were almost unheard of in 1959. The first such machine, the TR-14A Diesel from Wagner Tractor of Portland, Oregon, was tested at the University of Nebraska in June 1959. The TR-14A was a descendant of the earlier, non-articulated Wagner TR-9. These big, heavy, powerful movers stemmed originally from aircraft tugs. Powered by a six-cylinder, 148-hp Cummins engine, the TR-14A weighed a massive 21,050 pounds (9,473 kg) but could pull 10,749 pounds (4,837 kg).

At about the same time, articulated wheel loaders came on the scene. A firm named Scoopmobile made the first in 1952. Not many of these were delivered until later in the decade due to problems with the hinge. It is likely that this type of loader was related to the two-wheel tractor/scraper developed by LeTourneau of Peoria, Illinois, in the late 1930s, which incorporated hydraulic articulated steering and a heavy-duty center hinge.

1930s Massey-Harris GP brochure
Launched in 1936, Massey-Harris's 15/22 GP had "Four Wheel Drive! Balanced Traction! Flexibility!" according to this brochure. Sadly, the GP wasted too much of its four-cylinder Hercules engine's power getting power to the ground, and the machine never lived up to its promise.

1920s Allis-Chalmers Duplex

The amazing Duplex combined two 6/12 tractors mounted back to back, offering farmers a pioneering tandem tractor for heavy-duty chores. The transmission controls of the two tractors were joined together, giving the machine a crude form of four-wheel drive. The Duplex was years ahead of its time—too far ahead, in fact, and few were built or sold.

1960s Fordson Major Doe Triple D Conversion

In the quest for more power in the 1950s and 1960s, some handy farmers and intrepid firms combined two tractors into one. Two Fordson Majors were joined to make this articulated four-wheel-drive tractor before the days when production machines available. Owner: Jon Hooper.

1920s Fitch Four Drive brochure

The Four Drive Tractor Company of Big Rapids, Michigan, was one of several firms to pioneer four-wheel drive. As this brochure noted, "The drive on all FOUR of the wheels adds another pair of PULLERS instead of something to propel." The Fitch Four Drive debuted in 1916 and was built into the 1930s. It was powered by a Climax four and rated at 20/36 hp.

1968 Deere-Wagner WA-17

In 1968–1969, Deere marketed the 225-hp WA-14 and 280-hp turbo WA-17 four-wheel-drive articulated tractors made by the Wagner Tractor Company of Portland, Oregon. The WA-14 was powered by a 178-drawbar-hp Cummins N 855; the WA-17 by a 220-drawbar-hp Cummins NT 855.

John Deere Model 8010/8020

Deere & Company of Moline, Illinois

John Deere's Model 8010 sparked amazement when it was introduced at Deere & Company's field day in Marshalltown, Iowa, in 1959. The 8010 was huge: 20 feet long, 8 feet wide, and 8 feet tall (600x240x240 cm). It weighed 20,000 pounds (9,000 kg) without ballast and 24,000 pounds (10,800 kg) with liquid in tires that were nearly 6 feet (180 cm) high. And it was the first Deere in forty years to have more than two cylinders in its engine.

The 8010 boasted a six-cylinder GM 6-71 engine of 215 hp at a time when no other Deere tractor had yet exceeded 80 hp. It also had a nine-speed transmission while the most any other Deere had was a six-speed. Instead of mechanical brakes like every other Deere, the 8010 had air brakes.

1959 Deere 8010 brochure

Besides seeing what was being done with four-wheel-drive aircraft movers and end loaders, Deere was prompted to experiment with the 8010 when some northern Quebec farmers replaced the tracks on their John Deere Model B-Lindeman crawlers with balloon tires to provide increased floatation over muskeg soil. The converted B-Lindemans were steered like present-day skid-steer machines. Previously, Deere engineers experimented with a Model R tractor with a 108-hp GM diesel. This tractor was shown to a limited number of Great Plains farmers who loved the doubled horsepower, suggesting to Deere that big power would sell.

Unfortunately, the 8010 did not sell as expected. A total of 100 production models were built between 1960 and 1961. Only one 8010 was sold in 1960; it took until 1965 to sell the rest. The 8010 had some problems with the engine and transmission, and almost all of the 100 production models were recalled and rebuilt into the Model 8020, after which they were returned to their owners. Deere does not have records on about 15 of the 100 production 8010s, so they may not have been rebuilt.

The question is why such an advanced tractor sold so poorly? Did not the farmers that tried the 108-hp R ask for more power? Was not the reception at the Marshalltown unveiling of the 8010 positive? Even with the rebuilding into 8020s, one big problem remained: price. The cost for an 8020 with three-point hitch was $30,000—a chunk of change in the early 1960s considering that the 4020, Deere's next largest tractor, cost a mere $10,000.

Some of those who did farm with the 8010/8020 liked them and put them to good use. Others said the tractor was a dog; its power didn't live up to claims. Probably due in large part to the peculiarities of the two-cycle GM engine, farmers used to the pleasant exhaust note of the new Deere four- and six-cylinder engines didn't run the "Jimmy" hard enough. One owner reported that to operate an 8010/8020 properly, you had to slam your hand in the cab door about three times. Then, when good and mad, you were in the proper frame of mind to operate the engine.

1959 John Deere 8010

While other tractor makers were breaking the 100-hp barrier, Deere was breaking the 200-hp line. The big 8010 was one of the first articulated four-wheel-drive tractors and the first Deere in forty years to have more than two cylinders. The 10-ton (9,000-kg) 8010 had six cylinders in its GM 6-71 super-charged, two-cycle diesel rated at 215 hp. The 8010 shown here is the first one made, serial number 1000. It was never sold as a new tractor, but was used by Deere for tests and demonstrations. All of the 99 other 8010s were converted by Deere to 8020s. The Marshalltown, Iowa, dealer sold this 8010 for Deere to Walter Keller. At the time, it had a cracked head and a broken gear in the transmission. Repairs have been made, and the tractor is restored to new condition. Walter Keller is in the cockpit.

CHAPTER 6

The Modern Era, 1960–Present

"Only one in twenty Americans lived on a farm in 1980 as opposed to 1940 when one in four was a member of a farm family."
—Implement & Tractor, 1980

JOHN DEERE INTRODUCES

THE NEW THOROUGHBREDS OF POWER

21st CENTURY TECHNOLOGY TODAY

1990s John Deere 6900
Main photo: *Deere launched its new 6000 Series in 1992, adding the six-cylinder 6900 in 1994. This thoroughly modern tractor boasted 130 hp.*

1994 John Deere 8000 Series brochure
Above: *"The New Thoroughbreds of Power" brochure announced Deere's 8000 Series in 1994.*

Where Have All the Tractors Gone?

At its peak in the twentieth century, the farm tractor industry offered as many as 200 unique tractor brands. At the start of the twenty-first century, there were four tractor companies with American names: AGCO, CNH Global (Case–New Holland), Caterpillar, and Deere. Not all of these were American-owned, and a great number of the tractors from these manufacturers were made overseas.

Change has always been the rule in the tractor industry: mechanical evolutions and revolutions; corporate mergers, consolidations, and failures. But as with the ever-increasing pace of life, the changes today seem to be accelerated and more earth shattering than any that have come before.

1969 Ford 9000 and 1953 Ford Model NAA Jubilee
Two generations of tractors separated by only a decade—but oh, what changes and improvements! The 9000's six-cylinder, 401-ci (6,568-cc), turbocharged diesel produced 131 hp. Its transmission offered sixteen speeds forward with partial range power-shifting. Owner: Dean Simmons.

1962 Fordson Super Major

Left: *Until 1964, Ford of England made a separate line of tractors under the Fordson banner. The Super Major was a 5,500-pound (2,475-kg) tractor in the 60-hp class. The transmission was a three-speed with a high-low auxiliary. It had a three-point hitch with draft control. Owner: Dean Simmons.*

1964 John Deere 1010 Grove-Orchard

The 35-hp 1010 was the smallest of the New Generation Deere tractors. This is one of seventy-two Orchard 1010s made and one of only sixty-three with a gas engine. Owner: the Keller family.

John Deere 4020
Deere & Company of Moline, Illinois

John Deere's 4020 is one of the most significant tractors since the Ford-Ferguson 9N. It has been called the most copied tractor in history and was Deere's largest seller since the two-cylinder B.

The 4020 was actually an improved version of the original 4010. Introduced in late 1960, the 4010 was at the time the largest of Deere's New Generation tractors. The 4010's major features were an ergonomic operator platform with a seat designed by an orthopedic doctor, a lower-link draft-control three-point hitch, and a central hydraulic system.

The hydraulic system had the greatest influence on the design of future tractors. The system's heart was a variable-displacement pump driven from the front of the crankshaft. The pump supplied pressure for raising implements, both three-point and remote cylinders, the power steering (there was no mechanical connection to the front wheels), power brakes, and the differential lock.

1971 John Deere 4000 Low Profile
The 4000 was a lower-priced version of the 4020. Only forty-six were made. Owner: the Keller family.

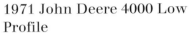

Built from 1963 to 1972, the 4020 offered an optional eight-speed power-shift transmission with four reverse ratios. The tractor was made in row-crop, high-crop, and standard-tread versions. It was available with gasoline, LPG, and diesel engines, although the diesel was the most popular.

The diesel version featured a six-cylinder, 404-ci (6,618-cc) engine. The gasoline and LPG engines displaced 340 ci (5,569 cc). Rated engine speed was 2,200 rpm for all versions. The diesel version developed 91 hp in its Nebraska test. The gasoline version produced 88 hp whereas the LPG version made 90 hp.

The 4020 was significant for another reason as well: It was the first John Deere two-wheel-drive tractor to bear a price tag of more than $10,000.

1962 John Deere 4010 HC LP
This was one of only twenty-three Hi-Crop 4010s powered by LPG; most were delivered to sugar-cane country in Louisiana. The 4010 had a 380-ci (6,224-cc), six-cylinder engine. It was also available as a diesel or a gas tractor, and in standard-tread, tricycle, or Hi-Crop configurations. Owner: the Keller family.

1969 John Deere 4020 brochure

Right: *Deere's 4020 was one of the most significant tractors in history. Introduced in late 1960, the 4020 was an improved version of the 4010. The 91-hp diesel version featured a six-cylinder, 404-ci (6,618-cc) engine. The 88-hp gasoline and 90-hp LPG engines displaced 340 ci (5,569 cc).*

1972 John Deere 4030

Below: *The 4030 used a 329-ci (5,389-cc), six-cylinder engine of 80 hp. This working tractor is owned by Glen Braun. "Lady" guards the tractor.*

Farmall 806
International Harvester Company of Chicago

International Harvester's answer to the Deere 4020 was the row-crop Farmall 806 and standard-tread International 806. The 806 featured a central hydraulic system similar to that pioneered by Deere, although the I-H system had separate circuits for steering, brakes, and implements.

The Farmall 806 was billed at the time as the world's most powerful all-purpose tractor at more than 90 hp. It recorded 94.93 PTO hp during its Nebraska test compared to the Deere 4020's 91.17 hp. The 806 used a six-cylinder engine displacing 361 ci (5,913 cc) for the diesel version and 301 ci (4,930 cc) for the gasoline and LPG models. Rated engine speed was 2,400 rpm as opposed to 2,200 rpm for the 4020.

The 806 sported an eight-speed transmission with a two-speed Torque Amplifier for sixteen forward ratios and two in reverse. Front wheel assist was an option.

The 806 was built between 1963 and 1967. It was replaced by the 100-hp 856, which remained in production through 1971.

1967 International 806
Billed as the most powerful all-purpose tractor of its day, the 806 recorded just under 95 hp in its Nebraska test with it's six-cylinder diesel engine of 361 ci (5,913 cc). Gas and a 301-ci (4,930-cc) LPG version were also available, as were both narrow and wide front ends. An eight-speed transmission was used by all. A central hydraulic system, after the fashion of the Deere 4010/4020, powered the steering, lift, and brakes. The 806, one of the most popular Internationals, was produced from 1963 to 1967.

The 100-Plus Horsepower Magic

Horsepower is simply the rate of doing work. Since farm work seems to always need to be done at once, there is no substitute for horsepower. One can only wonder at the patience of the farmer of the 1920s with hundreds of acres to till with a one-bottom plow!

In 1960, the average farm tractor had 40 to 50 hp. Today, that is the average for compact tractors. The largest of the two-wheel-drive tractors today has more horsepower than some four-wheel drives of the 1960s. Toward the end of the century, however, tillage practices and high costs have led to a decline in horsepower from the peak in the late 1970s.

With its RD-8 crawler, Caterpillar was the first to break the 100-hp barrier, as recorded by the University of Nebraska in 1936. The 34,000-pound (15,300-kg) tractor featured a six-cylinder, 1,246-ci (20,409-cc) engine. Cletrac soon followed with both its FD diesel and FG gasoline crawlers.

The first two-wheel-drive, wheel-type tractor to exceed 100 brake hp was the 1962 Deere 5010. Its six-cylinder, 531-ci (8,698-cc) diesel made 121 PTO hp. To harness this two-wheel-drive power without excessive slippage, 24.5x32 rear tires were used with a burdened weight of more than 17,000 pounds (7,650 kg). Allis-Chalmers also entered a two-wheel-drive machine in 1963 at 103 horsepower, the Model D-21.

Once the 100-hp genie was out of the bottle, everyone wanted part of the magic. All the manufacturers jumped on the bandwagon. Turbochargers, intercoolers, and after-coolers became commonplace in the horsepower race. The Case 1031 and Minneapolis-Moline G-1000 entered in 1966 as two-wheel-drive models. Four-wheel drives came into the picture in 1959 with the Deere 8010, 1962 International 4300, and 1964 Case 1200. Deere's 8010 was also the first farm tractor to exceed 200 hp.

The horsepower race was led by the Big Bud firm of Havre, Montana, which started constructing tractors in 1968. Big Bud's 1976 KT-450 had a 450-hp Cummins engine. The Big Bud 1978 model used on some California and other Western farms boasted a sixteen-cylinder, 760-hp engine. The current champion, however, is the giant, 800-plus-hp, $1-million Caterpillar D11, although it is not much used on farms except for deep sub-soiling.

"When you drop your implement and hit the throttle, you expect instant power and fast acceleration."
—Steiger Tractors brochure, 1980s

1974 Oliver 1655 Diesel
The horsepower race was on! Both the gasoline and diesel versions of the six-cylinder Model 1655 were rated at 70 hp. The diesel displaced 283 ci (4,635 cc) while the gasoline type displaced 265 ci (4,341 cc).

1974 Massey-Ferguson 1080

Right: *Using a Perkins 381-ci (6,241-cc) diesel, the 1080 was in the 80-hp class. Operating weight was about 12,000 pounds (5,400 kg). A six-speed transmission with partial-range power-shift gave twelve forward speeds. Owner: Maurice Burston.*

1963 Minneapolis-Moline M-504

Above, both photos: *This diesel M-504 with four-wheel drive was one of only twenty-one made. The M-504 was only built in 1962 and 1963. Owner: the Keller family.*

1966 Case 1030 Comfort King

The 1030 was available as a general-purpose model as shown, or in a standard-tread (Western) model. It featured the 451-ci (7,387-cc) six and was Case's first tractor to exceed 100 hp. Owner: Jay Foxworthy.

1971 Minneapolis-Moline G-1050 LPG

The G-1050 has a 504-ci (8,255-cc) six-cylinder engine producing 110 hp. Owner: Roger Mohr.

Massey-Ferguson trio
Above: *Massey-Ferguson's 9240, 6150, and 65 grace a sunny hillside.*

1975 International 966 Hydro
Left: *The '66 Series International tractors were built from 1971 to 1976. The 966 was a 100-hp example; this one had the optional full Hydrostatic transmission. A 414-ci (6,781-cc), six-cylinder diesel was used. Turbocharging was optional.*

1973 Ford 8600
Using the naturally aspirated Ford 401-ci (6,568-cc) diesel, the 8600 was the successor to the 8000. Internal improvements brought power up to 110 hp, and a partial range power-shift gave sixteen forward speeds.

1984 Ford TW-35 FWD Diesel
This 171-hp tractor featured a 401-ci (6,568-cc) engine with turbocharger and intercooler. It had a working weight of about 10 tons (9,000 kg). Owner: the Bissen family.

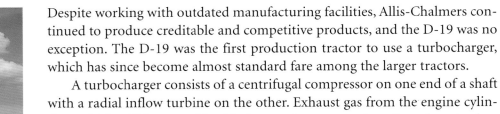

Allis-Chalmers D-19

Allis-Chalmers Company of Milwaukee, Wisconsin

1965 Allis-Chalmers ED-40

Above: *Similar to Allis's D-12, the ED-40 was built at A-C's factory in Essendine, England, for the British and Canadian market starting in 1963. It featured a 138-ci (2,260-cc), four-cylinder diesel engine and a four-speed transmission with a two-range auxiliary.*

1963 Allis-Chalmers D-19

Right: *The D-19 was built in Milwaukee, Wisconsin, from 1961 to 1964, with a few more than 10,000 constructed. They were available in gasoline, LPG, and diesel variations, but the diesel was noted for being the first production farm tractor to use a turbocharger. All engines were the A-C 262-ci (4,292-cc) six. Power was in the 65-hp class. Allis-Chalmers was one of the largest producers of turbochargers for aircraft during World War II. Owner: the Karg family.*

Despite working with outdated manufacturing facilities, Allis-Chalmers continued to produce creditable and competitive products, and the D-19 was no exception. The D-19 was the first production tractor to use a turbocharger, which has since become almost standard fare among the larger tractors.

A turbocharger consists of a centrifugal compressor on one end of a shaft with a radial inflow turbine on the other. Exhaust gas from the engine cylinders, which still has considerable heat and pressure when the exhaust valves open, drives the turbine to high speed. The turbine in turn powers the compressor via the common shaft. The compressor forces air into the intake manifold at two to three times atmospheric pressure. The harder the engine works, the more power there is in the exhaust to drive the compressor, which gives the engine more power to pull the load. Unlike superchargers, such as were used in the GM two-cycle diesels, the turbocharger consumes only power that would otherwise have been wasted in the exhaust. With a turbocharger, a small engine acts like one with a larger displacement, but consumes fuel like the smaller unit it is.

Swiss engineer Dr. Alfred J. Buchi invented the turbocharger in 1909. Nothing much came of it until World War II, however. General Electric, who had experience with water and steam turbines for driving generators, made turbochargers for most of the high-performance U.S. military aircraft, allowing them to retain their power as they went higher in altitude.

Production of Allis-Chalmers's turbo D-19 ran from 1961 to 1964, and only about 10,000 were built. It was available with LPG and gasoline engines, but only the diesel incorporated the turbocharger. All of the engines were six-cylinders displacing 262 ci (4,292 cc) and creating 65 PTO hp.

Following the D-19's successful application of the turbocharger, other tractors followed suit. Companies like M&W Gear Company sold aftermarket kits. By the end of the 1960s, all of the big U.S. tractor makers were offering turbocharged tractors.

226

1963 Allis-Chalmers D-21

Left: *The first Allis with more than 100 hp, the D-21 was rated at 119 hp. It was powered by a 426-ci (6,978-cc), six-cylinder diesel. After 1965, a turbocharged version was available with 127 hp. Owner: the Karg family.*

Factory Cabs: Comfort Comes to the Tractor

Although pioneered by Minneapolis-Moline with its UDLX of 1938, the factory-provided tractor cab did not become popular until the 1970s. In the original tractor designs, the operator sat as far back as possible in order to reach the implement controls. With the advent of hydraulic controls, the operator's seat was moved forward for a better, safer ride. Although the operator no longer had to reach external controls, many farmers did not like the idea of being inside when cultivating: They wanted to see the shovels going past the corn. As rear-mounted, three-point cultivators became more and more accepted, there was no way to actually watch the shovels. Looking backwards had to be done carefully, if at all, so the last resistance to the cab disappeared.

In the 1960s, aftermarket companies supplied cabs for the most popular tractor models. Many of these were only marginally satisfactory. Some actually increased the intensity of the noise and dust.

Cabs provided by the tractor companies were engineered for better results. Along with heating, air conditioning, and a nice radio came a comfortable seat—more like an office chair—with adjustments to fit the operator. Controls and displays were arranged for convenience. Positive-pressure fans kept the dust out and noise was so low that the radio could actually be used. In 1971, the University of Nebraska began reporting interior noise levels for all tractors tested.

1972 John Deere 4430
Above, both photos: *With a six-cylinder turbocharged engine of 404 ci (6,617 cc), the 4430 was a 126-hp tractor.*

1971 Allis-Chalmers One-Ninety XT Series III
Above and left: *The One-Ninety and One-Ninety XT were identical, except for the engine. The regular version used a 265-ci (4,341-cc) engine, while a 301-ci (4,930-cc) unit powered the XT. The One-Ninety was built from 1964 through 1973. A version of the XT Diesel used a turbocharger and made 93 drawbar hp. Owner: the Karg family.*

1973 Oliver 2255
Left: *Powered by a 573-ci (9,386-cc) Caterpillar V-8, this 17,000-pound (7,650-kg) tractor rated 147 hp, backed by an eighteen-speed transmission. Owner: Lee Miller.*

1988 Case-IH 956XL
Left: *The XL version of Case-IH's 956XL had a deluxe cab with air conditioning available.*

1980s Massey-Ferguson 3095
Below, both photos: *A Massey-Ferguson 3095 pulls a New Holland 865 baler.*

Transmissions: The Ratio Race

By 1960, multi-ratio transmissions were standard, and half-step power downshifts, like those pioneered by International Harvester's Torque Amplifier, were common. In 1957, Case introduced its Case-O-Matic torque-converter transmission, which had excellent load-starting ability. Oliver came out with a similar arrangement in 1958 with its Model 995 GM Lugmatic. In 1981, Steiger began using five-speed Allison (then a division of GM) torque-converter/power-shift transmissions with a two-speed auxiliary.

In 1959, Ford introduced the first all-power-shift transmission—the ten-speed Select-O-Speed. Deere offered a similar arrangement in 1963. Today, power-shift transmissions with up to eighteen speeds are available. The Caterpillar Challenger, for example, features a sixteen-speed forward unit (with nine reverse speeds) providing five shift modes, such as pulse shifting one gear at a time, continuous sequential shifting, preselected gear shifting, programmable up- and downshifting, or automatic shifting through gears ten through sixteen. In 1990, Ford offered a similar eighteen-speed unit called

1964 Minneapolis-Moline M-5
The famous M-5 was available in this LPG-powered high-crop version. Produced from 1960 to 1964, it was powered by a 336-ci (5,504-cc), four-cylinder engine with a five-speed transmission plus a two-speed power shift. The M-5 was also available in gasoline and diesel versions. Owner: the Keller family.

Ultra-Command. With it, the shift lever was moved forward and backward for forward and reverse in any gear selected. Gears were chosen by moving the shift lever left for lower gears, or right for higher speeds.

International Harvester introduced hydrostatic drives in 1969 on its 826 and 1026 models. These tractors also were equipped with manual high/low range shifters, but within either range, ground speed was infinitely variable at a constant engine speed. Versatile Manufacturing of Winnipeg, Manitoba, also offered a hydrostatic drive in 1977 on the Versatile Model 150.

In 1972, Oliver topped the ratio chart with eighteen forward speeds. This was accomplished with a six-speed conventional transmission and a three-speed power-shift auxiliary with under-, direct-, and overdrive. Allis-Chalmers took away those honors in 1973 with a twenty-speed arrangement on it 7030 and 7050. Massey-Ferguson offered a twenty-four-speed arrangement in 1978. Steiger boasted a twenty-four speed on the Tiger IV in 1984. Case-IH also had a twenty-four-speed unit in 1986, consisting of four gears and a six-speed power-shift. Deere offered a twenty-four-speed unit in 1989.

The 1987 Massey-Ferguson 3050 and 3060 came with thirty-two speeds accomplished with a four-speed power-shift, a gearbox with four synchronized gears plus a high/low range and a shuttle shift. In 1995, White also boasted of thirty-two speeds, but with an eight-speed manual gearbox and four-speed power-shift auxiliary. The all-time ratio champ, however, was the 1992 AGCO-Allis 8630 with forty-eight speeds: a sixteen-speed gearbox and a three-speed power-shift auxiliary.

1973 Ford 4000 Diesel
The 4000 was equipped with a three-cylinder, 201-ci (3,292-cc) engine, putting it in the 50-hp class. Weight was just under 5,000 pounds (2,250 kg), but that could be doubled with ballast. The eight-speed fixed-ratio transmission gave speeds from 1.5 to 17 mph (2.4–27 kmh). Owner: Dennis Burstin.

Since the 1960s, the tractor industry left behind distillate fuels and LPG. June 1968 saw the last LPG tractor test at the University of Nebraska.

The industry has also left behind gasoline. The International 284 was the last gasoline-powered tractor tested at the University of Nebraska, in 1978. Since that time, only diesel tractors have been available on the American market, with the exception of small lawn tractors. One exception is the "new" N Series Ford manufactured by N-Complete of Wilkinson, Indiana. According to N-Complete's Tom Armstrong, the Fords are re-manufactured using parts that meet new tractor specifications. The tractors also carry a warranty the same as a new machine.

One of the most profound changes went almost unnoticed, however: the disappearance of the tricycle front end. While the narrow, or tricycle front, remained in the catalogs for a time, they were rarely sold after 1973. Two things accounted for the demise of the narrow front: The use of chemicals largely eliminated the need for the cultivation of tall crops, and the advent of the front-end loader, which seemed to work so much better on a wide-front machine.

1966 Case 1030
This working 1030 was powering a grain auger filling rail cars for Farmer's Feed and Grain in Charles City, Iowa, in the summer of 1999. Wayne Bottloson was the operator.

1964 Minneapolis-Moline M-670
This early version of the M-670 was LPG powered and was one of the last with narrow-front. The M-670 was built from 1964 to 1970. Owner: Matt Ross.

The Proliferation of Articulated Four-Wheel-Drive Tractors

The trend toward larger, more powerful tractors continued in the 1960s. Steering limitations of powered front wheels led manufacturers to try both the skid-steer concept and the articulated four-wheel drive. Skid-steer soon lost out, and articulation became the chosen road.

1979 International 4586
Left: *The 4586 featured an 800-ci (13,104-cc) diesel and nine-speed transmission. It was built from 1976 through 1980.*

1980 John Deere 8440
Below: *The 8440 was powered by a 466-ci (7,633-cc) turbocharged and aftercooled diesel of 180 hp. Owner: Don Bray.*

1984 John Deere 8850

Above and right: *Powered by a turbocharged and aftercooled V-8 of 955 ci (15,643 cc), the 8850 was the top of Deere's line. Six headlights lit up the fields, so that the $120,000 machine could work day and night. Maximum power was 304 PTO hp. The big 8850 had a drawbar pull of 94 percent of its own weight. Owner: Ruth Schaefer.*

1982 John Deere 8450

Weighing nearly 15 tons (13,500 kg), the 8450 was a big four-wheel-drive articulated tractor. Its turbocharged and aftercooled six-cylinder engine developed 187 hp.

1964 Case 1200 Traction King

Case's first venture into four-wheel drive, the Traction King used four-wheel steering as well. Power was from a 461-ci (7,551-cc) diesel. Owner: John Thierer.

1979 Ford FW-30
Above and left: *Powered by a 903-ci (14,791-cc) Cummins V-8 diesel, the 32,000-pound (14,400-kg) FW-30 was a big machine for its time. Steiger made the FW Series tractors for Ford. Owner Dale Bissen added a turbocharger, upping power from 205 to 270 hp.*

1980 International 3588
The last of the new tractors from International Harvester before the merger with Case, the 2+2 Series tractors were the ultimate in their day. The 3588 had a 150-hp diesel.

1983 White Field Boss 4-225
Powered by a 636-ci (10,418-cc) turbocharged Caterpillar V-8, the 4-225 had a closed-center hydraulic system and eighteen-speed transmission. The cab featured a fourteen-channel digital monitor and air conditioning.

1971 White Plainsman A4T-1600
The articulated Plainsman began life under the Minneapolis-Moline banner. It was also sold as the Oliver 2655 before becoming the White Plainsman in 1970. The engine was an M-M 585-ci (9,582-cc) diesel of 143 hp. A ten-speed selective fixed-ratio transmission was used. It weighed 20,000 pounds (9,000 kg).

1982 International 6588
The International Harvester 2+2 design was introduced in January 1979. It was a unique approach to articulated four-wheel-drive tractors with the operator's cab behind the articulation point and the engine ahead. In fact, the engine was ahead of the front axle for weight distribution purposes. The large fuel tank was behind the front axle and acted as a sound shield between the engine and cab. Two sizes were available: the 130-hp 3388 and the 150-hp 3588. These were replaced by the 6388 and 6588 in the 1982 model year with transmission improvements, but with the same power. Two larger versions were in the works, the 210-hp 7288 and 235-hp 7488 when IH financial difficulties became insurmountable, and the Tenneco buyout was imminent. The buyout occurred in 1984 and the 2+2 tractors were dropped.

Versatile

Versatile Manufacturing Ltd. of Winnipeg, Manitoba

As a lad on his family's farm in Canora, Canada, Peter Pakosh was always fascinated by machinery. During the tough times of the Great Depression, he was forced to leave the farm and seek a job at Massey-Harris. When he was refused a transfer to Massey's design department in 1945, Pakosh began crafting his own machinery in his Toronto basement, building a grain auger that had fewer moving parts than existing models and would be less expensive to manufacture.

With the aid of his machinist brother-in-law, Roy Robinson, Pakosh started the Hydraulic Engineering Company in 1947, mortgaging everything the two owned and borrowing the egg money Pakosh's wife was saving for a fur coat. They branded their auger and new field sprayer the "Versatile," and incorporated in 1963 as Versatile Manufacturing Ltd.

In 1966, Pakosh launched his first four-wheel-drive tractor, which was the first mass-produced four-wheel drive, but it sold for about the same price as major makers' smaller two-wheel drives. Pakosh had earned a reputation for designing simple, straightforward machines that small farmers could afford to buy and could often maintain and repair themselves. Versatiles were mostly sold in Canada as well as in Minnesota, Montana, and South and North Dakota.

Pakosh sold Versatile in 1976 to Cornat Industries of Vancouver. In 1987, New Holland purchased the firm, but in 1999, Versatile was excluded from the CNH Global NV deal and is once again independent.

1977 Versatile 150
Introduced in 1977, the Model 150 was Versatile's revolutionary bi-directional tractor.

1980 Versatile 895
With an 855-ci (14,005-cc), super-charged and intercooled Cummins six-cylinder engine, Versatile's articulated 895 tractor produced 251 drawbar hp.

237

Steiger

Steiger Tractor Company of Fargo, North Dakota

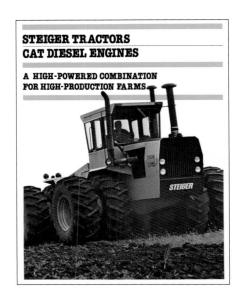

1980s Steiger brochure

By the 1980s, Steiger's lineup included the Cat-powered Bearcat III, Cougar III, Panther III, Panther 1000, and Tiger III.

Necessity was the mother of invention for brothers Maurice and Douglass Steiger. The duo who wanted a powerful four-wheel-drive tractor for their farm near Red Lake Falls, Minnesota, but no existing machine came close to meeting their needs. Starting with a Euclid earthmover, the brothers built their own tractor in the family dairy barn during the winter of 1957–1958.

The Steiger's 15,000-pound (6,750-kg) home-built machine was constructed from salvaged parts and powered by a 238-hp Detroit Diesel engine. Neighboring farmers liked the tractor and put down cash for a "Steiger" of their own. Some 125 tractors were eventually built on the farm.

Steiger Tractor was incorporated in 1969, with a factory in Fargo, North Dakota. That same year, Steiger's Series I Wildcat, Super Wildcat, Bearcat, Cougar, and Tiger machines hit the market. The firm was eventually contracted to build machines for a variety of the major makers, including Allis-Chalmers, Ford, and International Harvester.

By 1986, however, the ailing farm economy pushed Steiger into bankruptcy, and it was acquired by Tenneco. Steiger continued to build machines that were now painted Case red instead of Steiger green. By 2000, the firm was part of CNH Global NV and had built more than 50,000 four-wheel-drive tractors.

1982 Steiger Panther CP-1360

The 334-hp Panther was powered by a six-cylinder Caterpillar engine and built by the Steiger Tractor Company of Fargo, North Dakota. It boasted sixteen forward speeds. Weight was 35,000 pounds (15,750 kg).

1991 Case-IH 9280

In 1986, Tenneco took over the failing Steiger, folding it into its Case-IH line. By 1991, the 9280 was the largest tractor produced by the company. It used an 855-ci (14,005-cc) engine producing 344 hp with turbocharging and aftercooling.

1999 Case-IH Steiger 9330

AGCO: A Proud Heritage

On November 1, 1960, just before John F. Kennedy was elected president, the White Motor Corporation of Cleveland, Ohio, bought out the Oliver Corporation. The next year, the Canadian Cockshutt outfit was added to the White family, and Minneapolis-Moline was added in 1963. These brand names continued until 1969 when all were dropped in favor of the name White Field Boss.

In 1985, Allis-Chalmers was folded into Klockner-Humboldt-Deutz (KHD) of Germany and became Deutz-Allis. In 1990, Allis-Gleaner Company (AGCO) bought Deutz-Allis; in 1993, it acquired the White line. In 1993 and 1994, AGCO purchased the worldwide holdings of Massey-Ferguson, which by that time also included McConnell Tractors. Tractors produced under the AGCO banner now include AGCO Allis, AGCOSTAR, Massey-Ferguson, White, and Landini.

The heritage of these modern tractors goes all the way back to Hart-Parr. Many firsts were included in the designs that have led up to the present time: the first tractor factory; first production tractor model; first tractor advertising; first production kerosene-burning tractor; first multi-speed tractor transmission; first overhead-valve tractor engine; first production cultivating tractor; first articulated tractor; first tractor with starter and lights; first live PTO; first differential lock; first tractor to offer rubber tires as standard equipment; first LPG tractor engine; first tractor disk brakes; first tractor cab with radio; first four-valve-per-cylinder tractor engine.

It's a proud heritage.

1968 Minneapolis-Moline M-670

Available configured for gasoline, diesel, or LPG fuels, the M-670 used the M-M 336-ci (5,504-cc) engine block. All three versions were in the 70-hp class.

1963 Oliver 1600
Left, both photos: *Built in 1962 and 1963, the 1600 was available in LPG, gasoline, or diesel versions. This one was gasoline powered. Owner: Eldon Oleson, who worked at Oliver/White in Charles City, Iowa, for thirty-nine years.*

1960 Oliver 770
The 770 came out in 1958 and was an improvement over the previous Super 77. The six-cylinder engine gave about 55 hp in gasoline, diesel, or LPG versions. The gasoline engine is shown.

1970 Oliver 1755
The diesel engine in the 1755 offered 86 "certified" hp from 310 ci (5,078 cc). The gasoline version was slightly smaller at 283 ci (4,636 cc), but also gave 86 hp.

1975 Oliver 1755
This 1755 had the 310-ci (5,078-cc) diesel and the "Over-Under" power-shift along with the six-speed gearbox. Owner: David Preuhs.

1978 White 2-180 Series 3
Above and right: *The Field Boss 2-180 pulls a Model 605F Vermeer baler. The big White was powered by a Caterpillar V-8. Owner: Bruce Copper.*

1999 White 6124
The new White 6124 was in the 125-hp class.

1995 AGCO-Allis 8630

Above and left: 1992 Deutz-Allis 9130 FPA

1970 Case 1070 Agri King
Above and left: *The 451-ci (7,387-cc) Case 1070 was in the 100-hp category. The Model 1070 was sold from 1970 to 1978 with a six-cylinder engine and either an eight-speed manual or six-speed transmission with partial-range power-shift for twelve forward speeds.*

1976 Case 1570 Spirit of '76

Case built the 180-hp turbocharged 1570 in 1976 and 1977. During the U.S. Bicentennial year of 1976, special commemorative paint jobs were available. Owner: J. R. Gyger.

1960 Case 830 HC Comfort King

This high-clearance 830 had the Case-O-Matic transmission. The 830 was a 60-hp tractor. Owner: J. R. Gyger.

1973 International 1466 Turbo

Produced from 1971 to 1975, the 145-hp 1466 used a 436-ci (7,142-cc), six-cylinder diesel engine.

1973 International 1066

Built from 1971 to 1975, the 1066 was available with turbocharging and the fully hydrostatic transmission. Power, with the turbo, was 125 hp.

1979 International 784

The 784 featured a 65-hp, four-cylinder diesel of 246 ci (4,029 cc). The 84 Series tractors were built in IH's Doncaster, England, factory.

1993 Case-IH 7130

Built from 1988 to 1993, the 7130 featured a 505-ci (8,272-cc), six-cylinder diesel of 175 hp and an eighteen-speed power-shift transmission.

1982 International 5488

With its 466-ci (7,633-cc), six-cylinder diesel, the 5488 was the first International two-wheel drive in the 180-hp class, with a rating of 187 hp. Front wheel assist was later made available.

CNH Global NV: A Conglomerate Rich With History

The November 1999 merger of the Case Corporation and New Holland NV produced the world's leading maker of farm equipment. The company, which is expected to have sales of $12 billion, will be headquartered in Racine, Wisconsin. In 1844, when J. I. Case began manufacturing threshing machines in Rochester, Wisconsin, Rochester city fathers denied him water rights to the Fox River millrace. Case moved his operation twenty-five miles (40 km) east to Lake Michigan and the town of Racine. It was Rochester's loss.

Throughout its history, Case's management bounced from being either conservative or flamboyant. Despite this, the firm arrived in the 1960s in good shape, having just acquired the American Tractor Corporation in 1957, which put Case solidly into the crawler and construction-equipment businesses. By 1970, however, cash shortages forced Case to sell out to the giant conglomerate Tenneco of Houston, Texas. A bevy of technologically advanced tractors then issued from the Case factories. To expand its European foothold, Case acquired David Brown Tractors of England in 1972.

Meanwhile, the fierce competition in the tractor business was causing International Harvester insurmountable financial woes. IHC had been number one in tractor production until 1963, after which John Deere took the lead. IHC countered with thoroughly modern tractor designs, including new articulated four-wheel-drive models with the Control Center cab aft of the articulation point. In order to keep stockholders happy, however, IHC management failed to plow enough profits back into plants and technology. Even at that, stockholder unrest caused several management shakeups as dividends fell. Then, the United Auto Workers union struck for higher wages and increased benefits. The strike lasted six months, and IHC never recovered.

In 1985, Tenneco acquired IHC's Tractor and Implement Division, which was then folded into Case. The

1961 Fordson Dexta Diesel
A British-built tractor in the 30-hp class, the Dexta had a three-cylinder, 144-ci (2,359-cc), Perkins engine. The 1962 Super Dexta boasted a 152-ci (2,490-cc) engine. It was imported and sold in North America as the Ford 2000 Diesel. Owner: Dean Simmons.

1962 Ford 5000 Diesel
The 5000 Diesel was really a British Fordson Super Major in cream and blue paint and with different decals that was imported for sale in North America. Like the Super Major, it used a 220-ci (3,604-cc), four-cylinder Ford engine. Owner: Fred Bissen.

consolidation put Case-IH on a firm foundation for the future, and allowed the purchase of the ailing Steiger Tractor Company.

At the same time, Ford had acquired New Holland in 1986, and the Ford Tractor Division became Ford–New Holland. Ford also bought the Canadian Versatile Company. In 1994, Italy's Fiat Agri began a buyout of Ford–New Holland, which was completed in 1995. Since 1997, the line of tractors has been marketed worldwide under the New Holland brand. After eighty years of Ford tractors, the name disappeared from the field.

The new CNH Global NV company is 71 percent owned by Italian auto-making conglomerate Fiat. The U.S. Justice Department required some divestitures before approving the merger. Case must sell its ownership in Hay & Forage Industries, which was jointly held with AGCO. New Holland must sell off its Winnipeg, Manitoba, plant along with its Versatile and Genesis tractor lines. New Holland was allowed to keep its TV-140 bi-directional tractor, however. Its production is being moved to another facility.

1967 Ford 6000
Built from 1961 to 1967, the 6000 used the ten-speed power-shift Select-O-Speed transmission. Early models, painted red, had troubles with the transmission and were recalled. When reissued, the paint was changed to this blue and white. The 6000 was a 10,000-lb (4,500-kg), 60-hp tractor.

1972 Ford 8000
The 8000 was a 105-hp tractor with a 401-ci (6,568-cc), six-cylinder diesel. Working weight is about 15,000 pounds (6,750-kg). Owner: Floyd Dominique.

1972 Ford 7000
The 85-hp 7000 was equipped with a 256-ci (4,193-cc), four-cylinder diesel with turbocharger. Owner: Frank Bissen.

1976 Massey-Ferguson 1135

Above: *The 1135 had a 120-hp, 354-ci (5,799-cc) Perkins diesel with turbocharger. The tractor was two-wheel drive, but the reversed-tread front tires prevented sliding on wet ground. Owner: Rich Hollicky.*

1992 Ford 7840 Powerstar SL

Above, right: *One of a series of new Powerstar tractors for 1992, the 7840 used a 90-hp, 401-ci (6,568-cc) Genesis engine.*

1990s Massey-Ferguson 3655

An M-F 3655 works with a New Holland combine to harvest a wheatfield.

1990s Massey-Ferguson 8160

Deere and Caterpillar: Steering the Straight Course

Deere and Caterpillar have gone through this period virtually unscathed. They have neither acquired, nor have they been acquired. Both have had exemplary managements and outstanding products since early times.

The two Illinois-based companies have demonstrated an affinity for each other over the years. In 1928, as Deere was developing a combine for the market, Caterpillar offered to sell Deere its Western Harvester Division with its line of successful combines. The asking price of $1.25 million was more than Deere could handle after funding the development of the in-house design.

Seven years later, Caterpillar again approached Deere with an idea. Cat dealers were suffering in competition with Allis-Chalmers and International Harvester because they only had track-type tractors. Caterpillar purposed a joint dealer arrangement where both Deere wheeled tractors and Caterpillar crawlers would be sold. For further enticement, Caterpillar offered to *give* their combine line to Deere. Deere accepted and, in doing so, also obtained access to Caterpillar's foreign dealerships. The arrangement was maintained into the 1960s, but was substantially diminished after Deere began marketing its own crawlers.

Although Caterpillar had steered more into the heavy construction side of the business over the years, its management has always remembered the firm's agricultural roots. Special Ag versions of Cat's small- and mid-sized crawlers have been available. Then in 1987, the rubber-tracked Challengers made the scene, combining the advantages of both rubber-tired and track-type tractors with good highway transport speeds and high floatation over soft ground.

Both Deere and Caterpillar appear ready to tackle the new century and continue their long lineages.

1960 John Deere 3010
The 3010 sported a four-cylinder engine of 254 ci (4,161 cc). This one was gasoline-fueled, but diesel and LP were options.

1980 John Deere 2640

The Dubuque-built 2640 was a 70-hp, four-cylinder tractor. Displacement was 276 ci (4,521 cc).

1999 John Deere 8300T

Announced in 1996, production of the 8000T Series rubber-track models started in June 1997. For the first time since the 1960s, it was possible to purchase either a wheeled or tracked version of the same Deere tractor. This time, however, the tracks were rubber instead of steel link. The 8300T was rated at 205 hp.

1996 John Deere 6900

A John Deere 6900 pulls a Klaas Quadrant 1150 baler and sledge.

1996 John Deere 6400

The 6400 was the top-of-the-line four-cylinder Deere tractor. The 276-ci (4,521-cc) diesel produced 100 hp.

Caterpillar Challenger

Caterpillar Inc. of Peoria, Illinois

The first production farm tractor to combine the speed and mobility of a wheeled tractor with the floatation and traction of a crawler was the Caterpillar Challenger Model 65, introduced in 1987. From this single model, the rubber-tracked Challenger has grown into a full-line series covering a range of horsepowers and including three sizes of row-crop machines with adjustable track widths.

For 1999, conventional (non-row-crop) Challengers were in their fifth generation following the original Model 65. Included were the 310-hp 65E, 340-hp 75E, 370-hp 85E, and 410-hp 95E. The full power-shift transmission had ten speeds forward and two in reverse.

Row-crop models included the 175-hp 35, 200-hp 45, and 225-hp 55. The row-crop transmission had sixteen speeds forward with a shuttle reverse in nine speeds. All Challengers had hydrostatic differential steering controlled by a conventional steering wheel. All featured state-of-the-art cabs and electronics.

By 2000, both Case-IH and Deere had also entered the rubber-track market, and Claas KgaA of Germany began marketing Caterpillar's Challengers in Europe. Implements, such as spreaders and even combines, are now being made with rubber tracks. Watch for both smaller and larger tractors in the future to sport these new-technology treads.

1996 Caterpillar Challenger 45
The row-crop version of the rubber-tracked Challenger had 200 hp and a sixteen-speed transmission. Owner: John Yotter.

1992 Caterpillar Challenger 65B

Powered by Cat's 3306 turbocharged engine with 285 hp, the 65B had a ten-speed power-shift transmission and full hydrostatic steering. It is shown with a DMI 44.5-ft (13.4-m) field cultivator with which it can cover 30 acres (12 hectares) per hour. It is also equipped with a 500-gallon (1,650-l) chemical tank. Owner: Rich Hollicky. Hollicky's son, Scott, is an ag engineering student at the University of Illinois. Scott reported the Challenger was easy to drive, stayed straight on hillsides, and was comfortable for long days and nights in the field. They have had few repairs to the machine. When the alternator went out one spring Saturday night at 11:30, their dealer, Ziegler of Minneapolis, opened up at 1:30 Sunday morning to provide a new alternator, and the Challenger was back in the field by daybreak.

Index

About *the* Authors

Robert N. Pripps was born in 1932 on a small farm in northern Wisconsin. Besides farming, his father did local road building and maintenance with a JT crawler and Russell grader. Hard times during the Great Depression put an end to both the farming and road construction, and Bob's dad went to work as a Wisconsin Conservation Department Forest Ranger, a job that he held for the next thirty-five years. Living at the Ranger Station, Bob was always exposed to trucks and tractors. His first driving experience came at age nine when he disked a fire lane with an Allis-Chalmers crawler. During summers throughout World War II, Bob worked on neighboring farms, earning the opportunity to drive one farmer's new Farmall H at age eleven.

Bob's curiousity with things mechanical almost cost him his life at age twelve. He got a mitten caught in the power takeoff of a Gallion road grader. He extricated himself from the machine before it killed him, but the encounter cost him his right thumb.

When he was fourteen, another life-changing event occurred: His best friend's father bought the first Ford-Ferguson tractor in the area. Abject envy is not a pretty thing, but that's what reigned in Bob's heart. It was in no way sated until Bob got his own 2N at age fifty.

Bob went to high school in Eagle River, Wisconsin, earning his private pilot's license by the time he graduated in 1950. His missing right thumb kept him from military service, so he attended Parks Air College to study engineering and receive a commercial pilot's license and multi-engine rating. To help support himself, Bob took a night job at McDonnel Aircraft in St. Louis. Marriage and family responsibilities soon made the job a priority and schooling secondary. Bob became a flight test engineer on the RF-101 Voodoo while continuing night and correspondence school. Subsequent jobs in test engineering included Atlas missile base activation for General Dynamics and jet engine starter and constant speed drive testing for the Sundstrand Corporation.

After seventeen years of part-time classes, Bob graduated from college in 1969 with a Bachelor of Science in Marketing. Bob also held a certificate in Aeronautical Engineering by that time. He then served as the marketing manager for Sundstrand's Dayton, Ohio, office, retiring at fifty-five.

Along the way, Bob inherited thirty acres of maple forest that were part of the Wisconsin farm on which he was born. That's when he found justification for the Ford-Ferguson 2N that helps with harvesting sap for maple syrup. He later added a 1948 John Deere Model B, 1958 John Deere 440C, and 1962 Massey-Ferguson 85 to the farm.

After retiring, Bob wrote a book on his favorite tractor, the Ford. The book was published in 1990, teaming Bob with renowned English automotive photographer Andrew Morland. Since then, Bob and Andrew have collaborated on eleven books on classic tractors, and Bob has also authored five other tractor titles on his own.

Bob and his wife Janice now live in northern Wisconsin, almost within sight of the original homestead. Besides steady work on books, Bob and one of his three sons make about 150 gallons of maple syrup each spring.

Andrew Morland was educated in Great Britain. He completed one year at Taunton College of Art in Somerset and then three years at London College of Printing studying photography. He has worked since graduation as a freelance photojournalist, traveling throughout Europe and North America. His work has been published in numerous magazines and books; among his published books are *Classic Tractors of the World*, *Vintage Ford Tractors*, and *The Big Book of Caterpillar*, all published by Voyageur Press. His interests include tractors, machinery, old motorbikes, and cars. He lives in a thatched cottage in Somerset, Great Britain, that was built in the 1680s. He is married and has one daughter.